Real-time Adaptive Concepts in Acoustics

Real-time Adaptive Concepts in Acoustics

Blind Signal Separation and Multichannel Echo Cancellation

by

Daniel W.E. Schobben, Ph. D.
Philips Research Laboratories

SPRINGER-SCIENCE+BUSINESS MEDIA, B.V.

A C.I.P. Catalogue record for this book is available from the Library of Congress.

ISBN 978-94-010-3857-7 ISBN 978-94-010-0812-9 (eBook)
DOI 10.1007/ 978-94-010-0812-9

Additional material to this book can be downloaded from http://extras.springer.com

Printed on acid-free paper

Contents

List of Figures

Preface

This book deals with two topics; blind signal separation and multichannel acoustic echo cancellation. Both have received a lot of interest in literature lately. For multichannel echo cancellation this interest is mainly technology driven by the mass introduction of multichannel audio communication systems and speech recognition systems. Pioneering work on blind signal separation for audio applications has been published less than a decade ago. The topic has been booming ever since for the obvious reasons explained below.

Blind Signal Separation

The problem that is solved by blind signal separation is a magical one. This is especially true for acoustical applications where microphones for example pickup a mixture of various speech signals which are contaminated by room acoustics. The goal is to retrieve the independent speech signals without having knowledge of either (properties of) the speech signals or the acoustical transfer functions. The numerous applications that can be based on this technology include voice control and teleconferencing.

In 1995 the paper by Bell and Sejnowski, 1995a was published which gave both a theoretical background and reported promising results when working on real-world audio signals. This is one of the early publications that motivated many researchers to join the quest for the holy grail that is called blind signal separation.

Echo Cancellation

Echo cancellation is a well known technique that was developed about three decades ago. Its application can be found in any product that has both a microphone and a loudspeaker, the purpose of the echo canceler being to cancel the contribution of the loudspeaker signal to the microphone signal. It has been widely used in many consumer products over the last decade now that this technology has become economic to implement. Just when most aspects of echo cancellation have been studied intensively a new problem arises with the introduction of multichannel sound systems; that of multichannel echo cancellation.

DANIEL SCHOBBEN

I

REAL-TIME ADAPTIVE CONCEPTS IN ACOUSTICS: AN INTRODUCTION

Chapter 1

INTRODUCTION

Both topics that are covered in this book, blind signal separation and echo cancellation, are treated aiming at acoustical applications. This chapter gives a survey of these and other techniques along with related adaptive array concepts. Several applications are discussed that can be based on these techniques. Background information on room acoustics, speech and microphone setups is presented. Also the objectives of the work in this book are discussed and an overview of the remaining chapters is given. The chapter concludes with describing the mathematical notation that is used throughout.

1.1. Motivation

The steady development of cheap computing power over the last decade resulted in a renewed interest in acoustic signal processing theory that is now feasible for real-time operation. Due to the high computational cost, many acoustic signal processing algorithms have been restricted to process the data from one microphone only. Current technology now enables the integration of multiple algorithms using data from multiple microphones and/or loudspeakers running in real-time on a single digital signal processing core. Application areas include speech enhancement, teleconferencing and voiced controlled machinery. Typical signal processing algorithms that are required in these applications are blind signal separation, echo cancellation, noise reduction and dereverberation.

Not only can multiple microphone approaches result in better signal quality in existing applications but they also have strong potential towards new applications that rely on spatio-temporal filtering. Research is required to develop multichannel algorithms and to enable the integration of such algorithms that work on multiple microphone signals. Applying a traditional single channel algorithm independently to the signals of multiple microphones can result in a poor performance as these signals are not independent and should therefore not be treated independently. In addition, when multiple algorithms are applied, the operation of the individual algorithms depend upon each other. New comprehensive joint algorithms are required to obtain a good overall performance for real-world signals. These algorithms must be able to run in real-time. The work presented in this book is motivated by these observations.

1.2. Transparent Communication

Transparent communication refers to distant communication where the perceived sound and vision is transparent in the sense that the people involved appear to be sitting at the same table. Transparent audio communication consists of a recording part and a sound reproduction part. The sound reproduction part is beyond the scope of this book. The recording part ideally produces signals that are free of reverberation, noise and acoustic echoes. Also, it is desirable to separate speech signals that are coming from multiple speakers that are speaking simultaneously. Note that 'speakers' is used to indicate live speakers throughout, not loudspeakers. Good acoustic signal separation can often be achieved by using headphones for sound reproduction and close talking microphones for signal recording. Usually it is more desirable to have handsfree communication however. Signal processing is therefore required to achieve transparent communication. Application areas include teleconferencing, hands-free telephony and hearing aids.

Teleconferencing is a more general application of transparent communication in which one group of people talks to another group of people at a remote location. The remote speech is reproduced in the local room using loudspeakers. The speech in the local room is recorded using an array of microphones and transmitted to the remote location. Figure 1.1 depicts the teleconferencing setup, in which L_1 loudspeakers reproduce the far-end speech and L_2 microphones pick up a mixture of the near-end speech together with far-end speech contributions. A multichannel teleconferencing audio system provides a more realistic sound image than a single-channel system does. In practice, the reproduced sound image will usually not be the same as the original one although it can capture most of the properties that are important to the listener. In a multichannel system this can be achieved for example by reproducing the various speech signals through independent loudspeakers. With the additional spatial information listeners can more easily determine which speaker is speaking. Moreover, the speech of the speaker of interest is more intelligible to the listener due to the spatial discrimination that is exploited by human perception.

The near end signals that are picked up by the microphones require adaptive signal processing prior to transmission to the far end. Without this processing, the signals are degraded due to

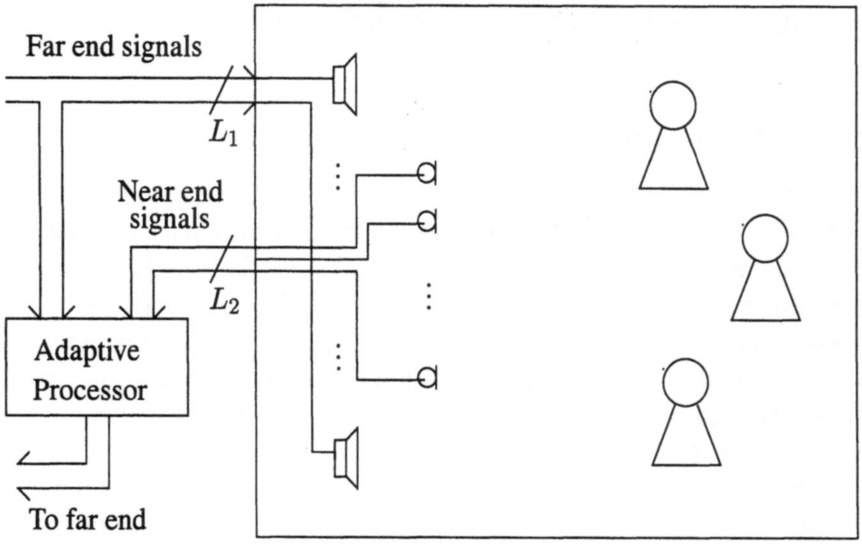

Figure 1.1. Teleconferencing setup

- Reproduced far-end signals which propagate towards the microphones and generate acoustic echoes

- The presence of several simultaneous speech signals causing the microphones to pick up an acoustic mixture instead of the individual speech signals

- Background noise which reduces the signal to noise ratio of the microphone signals

- Acoustic reverberation which affects the speech signals. Strong reverberation as in churches for example can render the recorded speech unintelligible. Also, in average office spaces or living rooms the intelligibility will decrease significantly however.

These degradations become worse and worse when people are not speaking close to their microphones. Signal processing algorithms are therefore required to undo these degradations. Multichannel signal processing algorithms have a large potential here as they can rely on the differences among the various acoustical transfer functions from a source to the multiple microphones.

An additional advantage of obtaining clean separated speech signals is that they can be coded much more efficiently for transmission over a channel of limited bandwidth. Besides the clean speech, information on the location of the speakers may also be transmitted as this is not contained in the separated speech signals. Such information is essential to make audio communication truly transparent.

1.3. Multichannel Signal Processing Concepts

Speech quality can be improved by applying signal processing to multiple microphone signals and far-end signals. Ideally the speech signals are separated, the acoustic echoes are canceled, the reverberation is eliminated and the background noise is suppressed to an inaudible level. In all of these cases, the signal processing involved is adaptive. This is required as the transfer functions from the speakers to the microphones are not known in advance and will also change in time. Movements of the speaker's head, for example, have a large impact on the transfer functions from this speaker to the microphones as the speaker is a directional sound source. Adaptive signal processing is also required in practical applications as the signals are non-stationary. A filter that is optimized for certain signals may result in poor performance when processing other signals.

Echo Cancellation

In applications such as hands-free telephony, a microphone picks up both the speech of the local speaker and the far-end speech that is reproduced by the loudspeaker inside the telephone. Besides degrading the recorded speech, this can result in acoustic feedback which is often perceived as a loud tone as the microphone signal is amplified and sent to the far-end where it again feeds into the remote microphone. An adaptive echo canceler can be used to eliminate the contribution of the loudspeaker signal to the microphone signal. In a teleconferencing setup, contributions of multiple loudspeaker signals to multiple microphone signals need to be canceled. In this book an algorithm is proposed for such a multichannel echo canceler. Also, the efficient implementation of large adaptive filters is treated in this book on which the multichannel echo canceler is based.

Blind Signal Separation

Humans can focus their attention on any one sound source out of a mixture. This was termed the 'cocktail party effect' in Cherry, 1953. This ability is due to the relations between the signals that are picked up by the left and the right ear, e.g. inter-aural differences in time and intensity. It is difficult, however, to understand speech that is recorded at a cocktail party with only one microphone, even for people with perfect hearing capabilities. People with hearing impairments readily have these problems when present at a cocktail party. Current audio systems cannot discern one sound from another in the way that humans can.

Blind Signal Separation (BSS) techniques have been proposed to recover source signals from microphone signals. These techniques are termed blind as the acoustic transfer functions from the sources to the microphones are unknown and there are no reference signals to compare the recovered source signals against. For recordings with one microphone, such techniques are based on speech properties, see for example Gu, 1992. When multiple microphone signals are available, BSS techniques are based on the fact that the source signals are independent of each other. They aim at recovering independent signals from the observed mixtures. These techniques typically require that there are as many microphones as there are sources. They have the potential to be more robust and they work for a wider class of signals, e.g. speech signals with mutually overlapping spectra. An overview of BSS using multiple microphones is given in this book and a new BSS algorithm based on second order statistics is proposed.

Dereverberation

When two people are far apart in a highly reverberant room they cannot have a conversation easily. This reverberation is due to strong echoes and decreases the intelligibility of recorded speech. Dereverberation is the process of removing the reverberation that is present in a signal. When performing dereverberation on a single channel recording, properties of the recorded signal are exploited, e.g. a cumulant-based approach was used in Jelonnek et al., 1997. In some applications, multiple microphone signals can be used so that dereverberation can be performed based on the differences between the transfer functions involved. Dereverberation is also called blind deconvolution in this context and is treated in Haykin, 2000a. Dereverberation is not discussed further in this book.

Noise Reduction

In acoustic applications noise can be classified as

- A physical noise source in the room, e.g. an overhead projection fan
- A diffuse noise field, e.g. noise in a car that is generated by wind
- Microphone/cable/amplifier noise, the microphone signals are contaminated with independent noise contributions.

The desired signals can be separated from physical noise sources using BSS. A diffuse noise field cannot be eliminated in this way as it consists of a superposition of an infinite number of noise sources. Besides the background noise that is present in any room, delocalized noise sources such as wind noise or the vibrating interior in a car also belong to this category. Techniques such as spectral subtraction can be used to eliminate diffuse noise, see Boll, 1979. Independent microphone noise may hamper the BSS algorithm severely. BSS systems do not have the potential to adequately deal with such independent recording noise. This is especially true when such noise contributions have energies that are comparable to those of the speech signals. There is no 'perfect' solution in this case. In the limiting case of relatively strong independent noise contributions the best solution for BSS is to perform no signal separation at all; separating the speech sources corresponds to mixing the independent noise sources. This solution is not favorable for the BSS algorithm as it tries to recover signals that are as independent as possible. Therefore preprocessing of the microphone signals is required to reduce these independent noise contributions. This can be achieved for instance by using linear prediction. The reduction of diffuse noise fields and independent microphone noise is not discussed further in this book.

1.4. Applications

All transparent communication applications are concerned with the distortion free recording of audio using one or more microphones and additional signal processing. The teleconferencing setup is a general and comprehensive application that involves multiple loudspeakers, multiple microphones and multiple speakers. Many other applications can be seen as special cases of teleconferencing. These are obtained by considering 0, 1 or multiple far-end signals that are reproduced by loudspeakers, 1 or multiple microphones and 1 or multiple speakers. Applications that are covered in this book are

- Acoustic Echo Cancellation (AEC) (1 far-end signal, 1 microphone, 1 speaker); an adaptive filter estimates the contribution of the far-end signal to the microphone signal. This contribution is subtracted from the microphone signal so that ideally the speakers' signal remains. An example application is hands-free telephony in which the loudspeaker signal feeds back into the microphone.

- Multichannel Acoustic Echo Cancellation (MC-AEC) (multiple far-end signals, 1 microphone, 1 speaker); using multiple single-channel AECs results in a slow convergence in the case where the far-end signals exhibit strong correlations. The MC-AEC can deal with this case however and can be used to prevent acoustic feedback when multiple loudspeakers reproduce multiple far-end signals, e.g. when someone listens to a stereo radio show and participates in the show by telephone.

  Note that in the case of only one loudspeaker and multiple microphones multiple single-channel AECs can be applied independently in a straightforward fashion. Similarly, canceling the contributions of multiple far-end signals to multiple microphone signals corresponds to applying multiple MC-AECs.

- Blind Signal Separation (BSS) (no far-end signals, multiple microphones, multiple speakers); multiple speakers that are talking simultaneously are recorded using multiple microphones. BSS is used to recover the individual speaker signals from these microphone signals. This can be used for example in a situation where multiple speakers are talking simultaneously into their own microphone and this microphone also picks up speech contributions from the other speakers. BSS can be used to remove such crosstalk.

- Combined BSS and MC-AEC (multiple far-end signals, multiple microphones, multiple speakers); the contribution of the loudspeaker signals to the microphones must ideally be eliminated and the local speakers' speech signals must be separated.

1.5. Acoustics and Speech

The signal processing algorithms that are involved in the applications mentioned above typically estimate room impulse responses (in the case of AEC) or inverse room impulse responses (in the case of BSS). It is therefore important to understand the properties of the acoustic environments.

Room Acoustics

The acoustic transmission from one point in space to another can be described by its impulse response. This impulse response is the sound pressure as a function of time due to an acoustic pulse emitted at the other position. Impulse responses are composed of contributions of the direct path and the many reflections by the walls. All these contributions have certain delays and attenuation due to propagation times and absorption by the walls. The absorption depends on the frequency of the sound and on the angle of incidence, see for example Cremer and Muller, 1982. Typically, the early reflections can be recognized by visual inspection of the impulse response. As the number of late reflections (reflections of reflections) is infinite, the tail of the impulse response has a dense structure. A room can be characterized by the reverberation time which is defined as the time that is needed for sound pressure level present in the room to decay by 60 dB. A short reverberation time (e.g. 100 ms) corresponds to a damped room. A long reverberation time corresponds to a concert hall (2-3 s) or a church (6 s). Rapidly decaying impulse responses correspond to short reverberation times while heavy tails correspond to long reverberation times. A typical office room has a reverberation time of approximately 500 ms, see Turner and Pretlove, 1991. Accurate impulse response estimates are required to be several thousands of coefficients long when sampled at a high sample frequency of for example 44.1 kHz. Although room impulse responses are of infinite length, longer estimation filters are not necessarily better; the tails of the room impulse responses become smaller and smaller in amplitude and the estimation errors grow as more filter coefficients need to be estimated. In many acoustical applications, filters with about one or two thousand coefficients give a good performance when operating at 44.1 kHz.

Microphones

Microphone responses depend on the frequency, the angle of incidence and the distance to the sound source. The type of microphone that is used in acoustic applications heavily influences the amount of reverberation and noise that is introduced into the recorded signal. Directional microphones capture sounds coming from the direction to which they are pointed. Reflections and background noise that come from other angles are suppressed to some extent. This results in low background noise and reverberation. Omni-directional microphones ideally have equal sensitivity for all angles of incidence. Signals from all angles are captured, but they are corrupted with strong reverberation and background noise. Cardiod microphones have a sensitivity that is in between that of directional and omni-directional microphones. Typically, they are sensitive to sounds "in front" of them and have low sensitivities to sounds coming "from behind". Low background noise and reverberation can also be achieved by using microphones which have a short effective range (e.g. microphones designed for recording vocals), see Clifford, 1986. The application determines the extent of prior knowledge available concerning the number of speakers and their (possible) positions with respect to the microphones. This restricts the types of microphones that may be appropriate. The choice of microphones can be of crucial importance.

Speech

From the theory of linear speech production, a short segment of speech can be modeled by an estimate of the vocal tract that is excited by the glottal excitation signal, see Owens, 1993. For voiced speech, this glottal excitation signal contains impulses corresponding to the pitch. For unvoiced speech, the glottal excitation signal resembles a noise signal. Voiced and unvoiced speech mainly cover the frequency range of 0.1-3.5 kHz and 0.1-7 kHz respectively. Speech signals are short-term stationary (over about 10 ms), see Schafer and Markel, 1979. Over longer time intervals, both the vocal tract and the type of glottal excitation change significantly resulting in highly non-stationary speech signals.

Speech intelligibility decreases with increasing levels of noise and reverberation. Also, hearing-impaired people suffer from attenuation and distortion of speech, see Plomp, 1978. Due to this distortion, the amplification of microphone signals containing noise and speech is insufficient to compensate for the hearing loss. Furthermore, speech coding algorithms, which can be used for example in teleconferencing, perform worse in noisy conditions. This leads to a further decrease in speech quality when the coded speech is to be transmitted over a channel of limited bandwidth.

The quality of speech is hard to qualify. Besides intelligibility, aspects like naturalness, brightness, signal processing artifacts and background noise play a part. A better and commonly used speech evaluation method is the *mean opinion score* which is obtained from averaging the opinions of many listeners. A disadvantage of this technique is that it is expensive and time consuming.

1.6. Outline

The objective of this work is to develop new multichannel algorithms for audio signal processing which can run in real-time. The main contributions in this book concern

- Single and multichannel acoustic echo cancellation
- Multichannel blind signal separation
- Combined blind signal separation and acoustic echo cancellation

All algorithms are developed in the frequency domain to ensure fast convergence and low computational complexity. An outline of the book is as follows;

Chapter 2, "Array Processing Techniques" presents an overview of existing array processing techniques for acoustic applications. This includes techniques such as beamforming, spectral estimation and noise cancellation.

Chapter 3, "Efficient Filtering using FFTs" provides a background of implementation details of frequency domain filters. The emphasis is on computational complexity and delay reduction for both single and multichannel frequency domain filters.

Chapter 4, "An Efficient Adaptive Filter Implementation" presents a computationally efficient frequency domain implementation of an acoustic echo canceler. The techniques used in this approach also apply to the algorithms in the following chapters.

Chapter 5, "Efficient Multichannel RLS" proposes a MC-AEC which is capable of canceling multiple echoes (e.g. signals that are generated by multiple loudspeakers) in a microphone recording. Multichannel decorrelation is used to prevent the correlations between the echo signals from deteriorating the overall system performance.

Chapter 6, "Blind Signal Separation, an overview" gives an overview of existing BSS algorithms and describes the basic ideas on which they are based. Indeterminacies of BSS based on second order statistics are considered in

Schobben and Sommen, 1999b. Implications of this work on the separation of real-world signals are presented in this chapter.

Chapter 7, "A New Blind Signal Separation Algorithm" proposes a new BSS algorithm based on second order statistics. Experiments involving recordings of live speakers in a room are performed to show the effectiveness of this approach. The computational complexity of the algorithm is computed indicating that the algorithm can be implemented in real-time. The work in this chapter is based on Schobben and Sommen, 1998a.

Chapter 8, "Joint Blind Signal Separation and Echo Cancellation" presents an extension of the proposed BSS algorithm for combined BSS and MC-AEC. The work in this chapter is published in Schobben and Sommen, 1999a.

Chapter 9, "Blind Signal Separation Algorithm Evaluation" presents methodologies for testing BSS algorithms using real and synthetic data. This work has been published in Schobben et al., 1999.

Chapter 10, "Conclusions" summarizes the main contributions of this work and gives the conclusions. Directions for future research are also given.

Appendix A, "Efficient Beamforming" gives an efficient implementation of a delay and sum beamformer in which the fractional delays are implemented using polyphase decomposition. This work has been published in Schobben and Sommen, 1996.

Appendix B, "On the Indeterminacies of Convolutive Blind Signal Separation based on Second Order Statistics" treats the mathematical and practical issue of whether second order statistics are sufficient or not for blind signal separation in audio applications.

Appendix C, "Computational Complexity of CoBliSS" gives an estimate of the computational complexity of the proposed blind signal separation algorithm. The properties of the matrices involved are exploited to obtain an efficient implementation. This work has been published in Schobben and Sommen, 1998b.

Appendix D, "Efficient computation of windowed FFTs" contains implementation details for the efficient implementation of frequency domain adaptive filters.

Appendix E, "Contents of the Compact Disc" describes the contents of the compact disc that is included with the book.

Appendix F, "Glossary" lists the abbreviations that are used.

1.7. Notation

It is required to understand the notation that follows in order to read the book. Besides this notation, a list of abbreviations included in the glossary for easy reference.

Time and frequency signals are denoted by lower case and upper case characters respectively. A character that denotes a vector is underlined. Superscripts denote the vector or matrix dimensions, a matrix with one superscript is square. As an example, a vector containing N samples of the signal x of which B samples are new for every increment of the block index k is denoted as

$$\underline{x}^N[kB] = \begin{pmatrix} x[kB - N + 1] \\ \vdots \\ x[kB] \end{pmatrix}$$

The l, m^{th} element of matrix A and the lth element of vector $\underline{B}$ are denoted as $(A)_{lm}$ and $(\underline{B})_l$ respectively. The $M \times M$ Fourier matrix $\mathcal{F}^M$ is defined as

$$(\mathcal{F}^M)_{kl} = e^{\frac{-2\jmath\pi(k-1)(l-1)}{M}} \quad \text{for } k, l = 1 \dots M$$

with $\jmath^2 = -1$. The $N \times N$ identity matrix and the $K \times L$ matrix containing zeros will be denoted by I^N and $0^{K,L}$ respectively. The $N \times N$ mirror matrix J^N has ones on its anti-diagonal and zeros elsewhere. Element-wise multiplication is denoted by $\otimes$, i.e.

$$\underline{C}^N = \underline{A}^N \otimes \underline{B}^N \Leftrightarrow (\underline{C}^N)_i = (\underline{A}^N)_i(\underline{B}^N)_i \quad \text{for } i = 1 \dots N.$$

The expectation operator gives the ensemble average of a variable and will be denoted by $E\{.\}$. The operator diag$\{.\}$ extracts the diagonal elements from a square matrix and puts them in a vector or converts a vector to a diagonal matrix depending on its argument. Also, A^*, A^T, A^H and A^{-1} denote complex conjugate, matrix transpose, hermitian transpose and matrix inverse, respectively.

Time indices of adaptive filter coefficients may be suppressed, their time dependency is readily understood. An update rule for the filters will therefore be denoted as $\underline{w}^N := \underline{w}^N + \dots$, where the colon that precedes the equality sign indicates that the right hand side of the equation is evaluated and assigned to the variable at the left hand side.

Often used symbols and operators are listed in Table 1.1 and Table 1.2.

f_s	Sample frequency (Hz)
$\underline{h}_{ij}$	Acoustic transfer function from source j to sensor i
$\jmath$	$\sqrt{-1}$
J^N	$N \times N$ mirror matrix
J	Number of speakers
I^N	$N \times N$ matrix identity
L	Length of the mixing filters or number of correlation-lags considered
L_1	Number of far-end signals
L_2	Number of microphones
N	Adaptive filter length
M	FFT size
$0^{K,L}$	$K \times L$ matrix containing zeros as its elements
$\underline{w}_{ij}^N$	Length N filter between sensor j and output i
$x[n]$	A time domain signal (denoted by lower case character)
$\underline{x}^N[kB]$	Time domain vector of length N (lower case underlined)
$\underline{X}^N[kB]$	Frequency domain vector of length N (upper case underlined)
$(A^{K,L})_{lm}$	l, m^{th} element of $K \times L$ matrix $A^{K,L}$
$(\underline{B}^N)_l$	l^{th} element of length N vector $\underline{B}^N$
α	forgetting factor for recursive power estimation
μ	adaptation parameter for adaptive filters

Table 1.1. List of symbols

$E\{.\}$	Expectation operator
$\otimes$	Element-wise multiplication
$\mathrm{diag}\{\underline{a}^N\}$	$N \times N$ diagonal matrix with $\underline{a}^N$ on its diagonal
$\mathrm{diag}\{A^N\}$	length N vector containing the diagonal elements of matrix A^N
A^*	Complex conjugate
A^T	Transpose
A^H	Complex conjugate transpose
A^{-1}	Matrix inverse
$A^{1/2}$	Matrix square root

Table 1.2. List of Operators

Chapter 2

ARRAY PROCESSING TECHNIQUES

Array signal processing provides an effective and versatile means of spatial and temporal-spatial filtering. Applications include radar, sonar, communications and geophysical exploration. This chapter gives an overview of array processing techniques for acoustic applications. As sound sources are wideband, temporal-spatial filtering is required which is generally more complex than spatial filtering alone. Array signal processing includes estimating the position of the sources, extracting these sources and inverting multichannel impulse responses. The sequel of the book will elaborate on the topics of blind signal separation and multichannel acoustic echo cancellation.

2.1. Beamforming

A beamformer performs temporal-spatial filtering to extract individual signals which may have overlapping frequency content but are generated by sources that are spatially apart. An introduction to beamforming is given in van Veen and Buckley, 1988. This section gives a survey of the basic principles of beamforming, ranging from traditional delay and sum beamforming to more sophisticated approaches.

Delay and Sum Beamforming

In a delay and sum beamformer the microphone signals are delayed, multiplied with weights and added. The propagation delays can be calculated from the positions of the microphones relative to the position of the desired sound source. The beamformer delays are chosen such that they compensate for the propagation delays. In this way, the desired source signal contributions in all microphone signals are added synchronously. Signals produced by unwanted sources at other spatial locations, recording noise and background noise are not added synchronously and hence a SNR improvement occurs. In a digital implementation of the delay and sum beamforming fractional time-delays can be used so that the beam can be steered accurately to the desired source more or less independent of the sample frequency. Considerations on the efficient implementation of the delay and sum beamformer using fractional time-delays are given in Appendix A.

Filter and Sum Beamforming

In Filter and sum beamforming, Finite Impulse Response filters (FIRs) are used to process each microphone signal. This allows for performing temporal-spatial filtering. A broadband beamformer can be designed such that it has a spatial sensitivity that is independent of the frequency of the signals van Gerven et al., 1995.

In practical applications, good parameterizations of the sources and the acoustic properties of the room are not available. Also, people move their heads or may walk through the room which results in a reduced performance when a non-adaptive beamformer is used. An adaptive beamformer is required to deal with such a dynamical environment. Two approaches for adaptive temporal-spatial beamforming are given next.

Constrained Beamforming

When there is partial knowledge of the acoustic environment, this knowledge can be incorporated in the adaptive beamforming algorithm using constraints. When only the position of the desired source is known for example, the beamformer is constrained to give a fixed response for signals coming from that location. All other degrees of freedom are adaptively chosen to minimize the output power of the beamformer and therefore reducing microphone noise, background noise and interfering sources. A multichannel FIR filter is used which is described as

$$y[n] = \sum_{l=0}^{N-1} \sum_{m=1}^{J} w_m[l] x_m[n-l] = \sum_{m=1}^{J} (\underline{w}_m^N)^T \underline{x}_m^N[n] = (\underline{w}^{JN})^T \underline{x}^{JN}[n]$$

with J the number of microphones, N the filter length, $x_m[n]$ the mth microphone signal, $y[n]$ the beamformer output,

$$\underline{w}_m^N = \begin{pmatrix} w_m[N-1] \\ \vdots \\ w_m[0] \end{pmatrix} \quad \text{and} \quad \underline{x}_m^N[n] = \begin{pmatrix} x_m[n-N+1] \\ \vdots \\ x_m[n] \end{pmatrix}.$$

The $\underline{w}^{JN}$ and $\underline{x}^{JN}[n]$ are obtained by stacking all $\underline{w}_m^N$ and $\underline{x}_m^N[n]$ respectively. The constrained beamformer objective is to minimize the beamformer output power while fulfilling the constraints

$$\min_{\underline{w}^{JN}} \frac{1}{2}(\underline{w}^{JN})^T R_x^{JN} \underline{w}^{JN} \qquad \text{subject to} \qquad (C^{JN,p})^T \underline{w}^{JN} = \underline{f}^p$$

with the correlation matrix

$$R_x^{JN} = E\{\underline{x}^{JN}[n](\underline{x}^{JN}[n])^T\}$$

and p the number of constraints that are imposed. The constraints are chosen such that the desired signal is preserved. When the desired signal comes from broadside (direction perpendicular to the array), the constraint matrix $C^{JN,p}$ reduces to a vector containing 1's and the gain vector $\underline{f}^p$ reduces to a scalar containing a number greater than 0. Now the array is guaranteed to have a gain equal to the number in $\underline{f}^p$ for signals that arrive synchronously at all microphones. The optimal solution can be found by defining a cost function that includes the beamformer output power and a penalty on deviations from the constraint

$$I(\underline{w}^{JN}) = \frac{1}{2}(\underline{w}^{JN})^T R_x^{JN} \underline{w}^{JN} + (\underline{\Lambda}^p)^T((C^{JN,p})^T \underline{w}^{JN} - \underline{f}^p)$$

with $\underline{\Lambda}^p$ a set of Lagrange multipliers. Setting the derivative of this cost function with respect to $\underline{w}^{JN}$ to zero

$$\frac{\delta I(\underline{w}^{JN})}{\delta \underline{w}^{JN}} = R_x^{JN} \underline{w}^{JN} + C^{JN,p} \underline{\Lambda}^p = 0$$

and solving for the Lagrange multipliers[1] yields the optimal solution, see Applebaum, 1976; Griffiths, 1969

$$\underline{w}_{\text{opt}}^{JN} = (R_x^{JN})^{-1} C^{JN,p} [(C^{JN,p})^T (R_x^{JN})^{-1} C^{JN,p}]^{-1} \underline{f}^p.$$

The derivative of the cost function can also be used in a gradient descent adaptive learning rule by replacing

$$\begin{aligned} R_x^{JN} \underline{w}^{JN} &= E\{\underline{x}^{JN}[n](\underline{x}^{JN}[n])^T\} \underline{w}^{JN} \\ &= E\{\underline{x}^{JN}[n] y[n]\} \end{aligned}$$

by its instantaneous estimate $\underline{x}^{JN}[n]y[n]$ as presented in Frost, 1972

$$\begin{aligned} \underline{w}^{JN} &:= \underline{w}^{JN} - \mu \frac{\delta I(\underline{w}^{JN})}{\delta \underline{w}^{JN}} \\ &\approx \underline{w}^{JN} - \mu(\underline{x}^{JN}[n]y[n] + C^{JN,p}\underline{\Lambda}^p) \\ &= P^{JN}\left(\underline{w}^{JN} - \mu \underline{x}^{JN}[n]y[n]\right) + \underline{F}^{JN} \end{aligned}$$

with

$$\begin{aligned} P^{JN} &= I^{JN} - C^{JN,p}[(C^{JN,p})^T C^{JN,p}]^{-1}(C^{JN,p})^T \\ \underline{F}^{JN} &= C^{JN,p}[(C^{JN,p})^T C^{JN,p}]^{-1}\underline{f}^p \end{aligned} \tag{2.1}$$

and μ the adaptation step-size. The Lagrange multipliers $\underline{\Lambda}^p$ are eliminated by demanding that the updated weight vector satisfies the constraints

$$(C^{JN,p})^T \underline{w}^{JN} = \underline{f}^p.$$

The projection matrix P^{JN} removes all components that are in the constraint subspace spanned by the columns of $C^{JN,p}$. These components are replaced by the fixed transformed constraint vector $\underline{F}^{JN}$.

Although theoretically appealing, there are several drawbacks to this approach. If the constraints are used to receive a signal from a certain spatial position with a constant temporal-spatial response over a wide frequency range, many constraints are needed. This limits the degrees of freedom for the adaptive output power minimization. Moreover, the projection matrix is very large for acoustic applications. For a 5 microphone array and filters with 200 taps for example, it is of size 1000×1000. Operations such as matrix multiplication are prohibited for such large matrices when real-time operation is required.

Side-lobe Canceler

The side-lobe canceler uses an approach that implicitly imposes constraints rather than projecting the outcome of an unconstrained algorithm onto the constraint subspace. In a preprocessing step a main beam and several auxiliary beams are formed. These are designed such that the target signal passes through the main beam but it is prevented from passing through the auxiliary beams. An

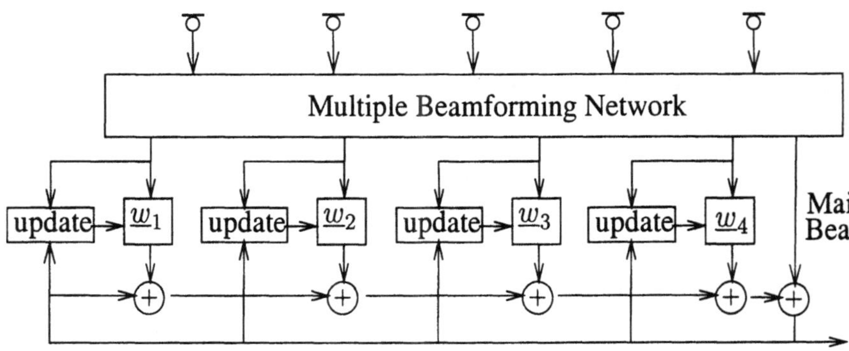

Figure 2.1. Side-lobe canceler

unconstrained adaptive algorithm can now be used to filter the auxiliary beams and add them to the main beam to minimize the output power. The side-lobe canceler is plotted in Figure 2.1. The fixed beamformer that generates the main beam and the auxiliary beams can be implemented using fractional delays as described in Appendix A or by using a multichannel FIR structure. Important in the design is that the auxiliary beams have sharp nulls at the position of the maximum of the main beam to prevent the target signal from leaking into the auxiliary beams. Fast convergence is achieved when all beams have sharp nulls at the positions where the other beams have their maximums so that there contributions can be adapted more or less independent of each other. In Figure 2.2 an example is plotted of the spatial sensitivity of the main beam in a five element array which is pointed to the angle of 0°. The spatial sensitivities of the auxiliary beams are plotted in Figure 2.3. The constraints are now incorpo-

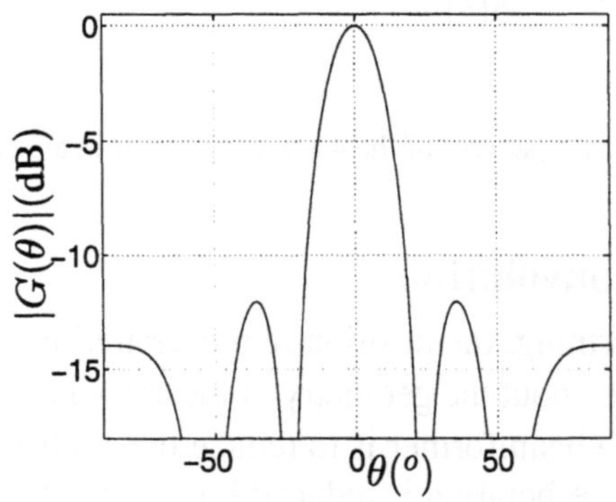

Figure 2.2. Beam-pattern of the main beam in a 5 element array

rated implicitly in the multiple beamforming network so that the update can be done using an unconstrained adaptive algorithm, see Griffiths and Jim, 1982. A side-lobe canceler can be implemented efficiently in the frequency domain using partitioned frequency domain adaptive filters as described in Joho and Moschytz, 1997. Partitioned frequency domain adaptive filters are described in for example Sommen, 1989. A straightforward implementation of the side-lobe canceler is not able to deal with the difficulties of real-world scenarios where microphones are placed in a small enclosure such as a room or a car. In such applications reflections generate copies of the signals that come from all directions. It is therefore insufficient to merely compensate for the time delays that are involved in the propagation of the desired and undesired signals to the microphone array so that more sophisticated beamformers are required as described in Affes and Grenier, 1997.

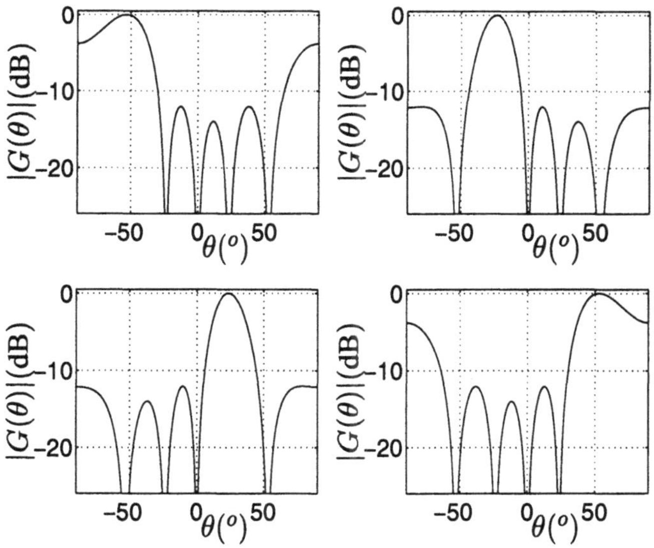

Figure 2.3. Beam-patterns of the 4 auxiliary beams in a 5 element array

2.2. Blind Deconvolution

In the case of beamforming, the position of the desired target is assumed to be known and knowledge about the geometry of the array is used to focus onto the target. The goal of the beamformer is to reduce the contributions of unwanted sources and noise. Reverberation is reduced however as the beamformer is not sensitive to signals that come from directions other than that of the target signal.

Classic deconvolution deals with recovering signals that are observed through an unknown system. An introduction to blind deconvolution can be found in Cadzow, 1996; Haykin, 1994. For single channel blind deconvolution assumptions are commonly made on the probability density functions (PDFs) of the sources. In Figure 2.4 a single channel blind deconvolution scheme is given. The non-linear element $g(.)$ is chosen in correspondence with the PDF of the

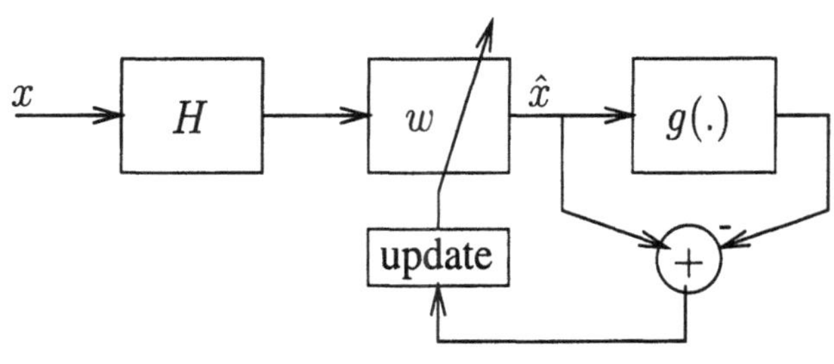

Figure 2.4. Single-channel blind deconvolution

source signals. The difference between the input and the output of the non-linearity is then used as the error signal that is minimized by the update. In telecommunications for example, binary sequences can be recovered by using a sign function for $g(.)$. When excited by any zero-mean random process, the output of such a non-linear element resembles that of a random binary sequence. Hence, to minimize the error signal, the adaptive filter is updated such that it recovers a binary output.

In multichannel blind deconvolution two aspects can be distinguished. One being the equalization of the channel that distorts the observed signal analogous to single channel blind deconvolution. To other one being the removal of crosstalk, e.g. the elimination of the contributions of all other sources but one in the observed signals. In an acoustic context, the former is termed dereverberation and the latter is termed Blind Signal Separation (BSS). An introduction to BSS is given in Chapter 6. The additional assumption that is generally made on the source signals to achieve BSS is that they are independent.

2.3. Detection of Multiple Signals

In beamforming and many other applications it is desirable to know how many sources are present and where they are located. Super-resolution techniques have been developed which are based on the eigenstructure of the covariance matrix of the microphone data, see for example Pillai, 1989. When the L_2 microphones pick up J narrow-band non-coherent sources corrupted with spatially and temporally independent noise, the microphone covariance matrix can be written as

$$R_{x}^{L_2} = E[\underline{x}^{L_2}(\underline{x}^{L_2})^H] = A^{L_2,J} R_s^J (A^{L_2,J})^H + \sigma^2 \mathrm{I}^{L_2}$$

with $A^{L_2,J}$ a complex transfer matrix describing the transfer function from the sources to the microphones. Complex notation is convenient here, as it allows for phase shifting by multiplying with scalars. Also, σ^2 is the microphone noise power and R_s^J is the source covariance matrix. Real data can be transformed to complex data using a Hilbert transform, details can be found in Appendix A of Compton, 1988. The eigenvalue decomposition of $R_x^{L_2}$ is

$$R_x^{L_2} = B^{L_2} \Lambda^{L_2} (B^{L_2})^H$$

with

$$\Lambda^{L_2} = \mathrm{diag}[\lambda_1 + \sigma^2, \lambda_2 + \sigma^2, \ldots, \lambda_J + \sigma^2, \sigma^2, \sigma^2, \ldots, \sigma^2]$$

and with $J < L_2$. In the noiseless case, this matrix contains $\lambda_1 \ldots \lambda_J$ on its diagonal due to the source signals and 0's for all other eigenvalues. The signal subspace is spanned by the eigenvectors corresponding to the non-zero eigenvalues. In the noisy case, the noise contributes equally to all eigenvalues. The noise subspace is spanned by the eigenvectors corresponding to the eigenvalues that are equal to σ^2. The remaining eigenvectors span a subspace in which both the source signals and also some noise resides. This subspace is termed the signal subspace. An interesting observation is that the noise power can readily be reduced by projecting the microphone data onto the signal subspace. The projection matrix can be constructed as in (2.1).

The detection of multiple signals can be done based on the array geometry. For a linear array for example, all microphones are positioned along a line and are spaced d apart. Consider the simplest case where there are no reflections due to the walls of a room. The columns of the transfer matrix $A^{L_2,J}$ are called the steering vectors and the lth column $\underline{a}^{L_2}(\omega_l, \theta_l)$ represents the transfer function from the lth source to the array

$$\underline{a}^{L_2}(\omega_l, \theta_l) = \left[e^{\jmath\omega_l\tau_1(\theta_l)} e^{\jmath\omega_l\tau_2(\theta_l)} \ldots e^{\jmath\omega_l\tau_{L_2}(\theta_l)} \right]^H$$

with $\tau_i(\theta_l) = (i-1)\frac{d}{c}\sin(\theta_l)$, c the propagation speed of sound through air and θ the direction of arrival relative to broadside. A signal classification algorithm termed the MUSIC algorithm is based on the observation that all columns of $A^{L_2,J}$ are perpendicular to the eigenvectors of the microphone signal covariance matrix that span the noise subspace, i.e.

$$(\underline{\beta}_j^{L_2})^H \underline{a}^{L_2}(\omega_l, \theta_l) = 0\,, J \leq j < L_2\,,\ 1 \leq l \leq J$$

where the eigenvector $\underline{\beta}_j^{L_2}$ is the jth column of B^{L_2}. The steering vectors can now be found by picking the peaks of a suitable function of all these inner-products see Pillai, 1989. Such a function can be chosen such that these peaks are extremely sharp and therefore very accurate. The resolution of signal detection using super-resolution techniques is therefore much higher than that of signal extraction when using the same array configuration. The detection of wideband sources is required in acoustic applications. Suitable approaches are extensions of the classical MUSIC algorithm as is described in Su and Morf, 1983; Wang and Kaveh, 1985. These approaches decompose the microphone data using Fourier transforms and perform MUSIC per frequency band.

2.4. Inverse Room Acoustics

In Section 1.5 some properties of room acoustics are discussed. In applications involving for example deconvolution and blind signal separation the inverses of room impulse responses are sought. Besides separation and dereverberation of signals, inverse room acoustics also applies to sound reproduction systems where the source signals are filtered by estimates of the acoustic transfer functions inverses prior to reproduction. This allows for perfect sound reproduction at a finite number of positions in space as described in Nelson et al., 1995.

Because a room impulse response generally is non-minimum phase, an exact inverse of a single channel acoustic transfer function does not exist. This problem is relaxed however when delay is introduced into the inverse system[2]. Perfect reception or reproduction of a single source is possible however using a multichannel approach as proved in the multiple-input/multiple-output inverse filtering theorem (MINT) Miyoshi and Kaneda, 1988. The multichannel approach does not require the introduction of delay. In this section, the recovering of multiple signals is considered and a distinction is made between separation and dereverberation as discussed in Section 2.2.

System Model

A blind deconvolution application is depicted in Figure 2.5. The microphone signals $x_1 \ldots x_{L_2}$ which contain contributions of all source signals $s_1 \ldots s_J$ are processed to obtain the output signals $y_1 \ldots y_J$. All transfer functions are assumed to be FIR filters. The transfer function from source s_i to microphone x_j is denoted $\underline{h}_{ji}^L$ and the acoustic transfer function from microphone signal x_i to output y_j is denoted $\underline{w}_{ji}^N$. The desired impulse response $\underline{d}_{ji}^{N+L-1}$ from

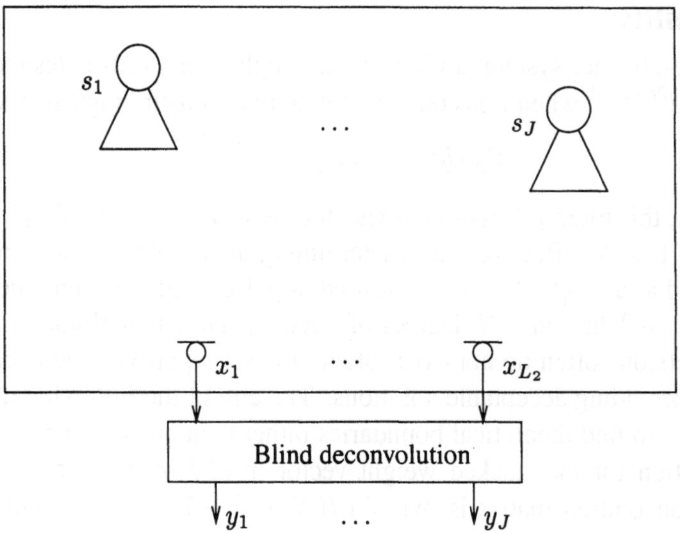

Figure 2.5. Blind deconvolution system

source s_i to output y_j is written as a product of a multichannel convolution matrix containing the acoustic impulse responses and a vector containing the deconvolution filters

$$\left(\begin{array}{ccc|c|ccc} \underline{h}_{1i}^L & & 0 & & \underline{h}_{L_2 i}^L & & 0 \\ & \ddots & & \cdots & & \ddots & \\ 0 & & \underline{h}_{1i}^L & & 0 & & \underline{h}_{L_2 i}^L \end{array}\right) \begin{pmatrix} \underline{w}_{j1}^N \\ \vdots \\ \underline{w}_{jL_2}^N \end{pmatrix} = \underline{d}_{ji}^{N+L-1}. \quad (2.2)$$

Introducing the vector to convolution-matrix operator

$$\mathcal{C}_N(\underline{h}_{ji}^L) = \begin{pmatrix} \underline{h}_{ji}^L & & 0 \\ & \ddots & \\ 0 & & \underline{h}_{ji}^L \end{pmatrix},$$

which has dimensions $N + L - 1 \times N$ and stacking (2.2) for $i = 1 \ldots J$ gives the desired response from all sources to output y_j

$$\begin{pmatrix} \mathcal{C}_N(\underline{h}_{11}^L) & \cdots & \mathcal{C}_N(\underline{h}_{L_2 1}^L) \\ \vdots & \ddots & \vdots \\ \mathcal{C}_N(\underline{h}_{1J}^L) & \cdots & \mathcal{C}_N(\underline{h}_{L_2 J}^L) \end{pmatrix} \begin{pmatrix} \underline{w}_{j1}^N \\ \vdots \\ \underline{w}_{jL_2}^N \end{pmatrix} = \begin{pmatrix} \underline{d}_{j1}^{N+L-1} \\ \vdots \\ \underline{d}_{jJ}^{N+L-1} \end{pmatrix}. \quad (2.3)$$

The stacked convolution matrix has dimensions $J(N + L - 1) \times L_2 N$.

Invertibility

In a single-channel system, a filter $\underline{w}^N$ is sought that yields a desired impulse response $\underline{d}^{N+L-1}$ when it is convolved with the acoustic impulse response $\underline{h}^L$

$$\mathcal{C}_N(\underline{h}^L)\underline{w}^N = \underline{d}^{N+L-1}$$

Obviously, this filter $\underline{w}^N$ does not exist in general as the matrix $\mathcal{C}_N(\underline{h}^L)$ is 'tall' $(N + L - 1 > N)$ effectively overdetermining the problem. This can be easily understood as a length $N + L - 1$ desired impulse response cannot be achieved as the filter $\underline{w}^N$ has only N degrees of freedom. Note that although in general exact inversions often do not exist, often practical approximations can still be obtained providing acceptable solutions. The aim of the following derivations is therefore to find theoretical boundaries rather than practical ones.

A solution for the stacked weight vector in (2.3) can be found when the stacked convolution matrix is 'wide' $(J(N + L - 1) \leq L_2 N)$ and full rank. The corresponding filter length required for perfect recovery of the sources is therefore

$$N \geq \frac{J}{L_2 - J}(L - 1).$$

Recovering the sources from a convolutive mixture $(L > 1)$ is therefore not possible when there are as many microphones as there are sources $(L_2 = J)$, no matter how many coefficients are used in the deconvolution filters.

This is an interesting observation as many researchers use FIR filters in their attempt to recover J sources from J microphones. Although exact inversion of the multichannel room impulse response is not possible in some cases, it is still possible that satisfying results are obtained. Optimal solutions can be found in least-squares sense within the restriction of the FIR filter structure. Depending on the signals, the room impulse responses and the length of the deconvolution filters, these solutions can be fully acceptable to the human ear.

Signal Separation

Although joint dereverberation and signal separation cannot be obtained when there are as many microphones as there are sources, it is interesting to know if it is possible to perform signal separation only. To simplify notation, the outputs are arranged so that the jth source is recovered at the jth output. For signal separation alone, there is no desired transfer function from source s_j to output y_j, i.e. signal separation is achieved when the recovered signals are filtered versions of the sources. This can be expressed by eliminating the jth row of convolution matrices from the stacked convolution matrix and removing the corresponding desired response vector $\underline{d}_{jj}^{N+L-1}$ in (2.3) for all j. The reduced stacked convolution matrix has dimensions $(J-1)(N+L-1) \times L_2N$. Separation is possible if this matrix is 'wide' ($(J-1)(N+L-1) \leq L_2N$). The length of the separation filters must therefore satisfy

$$N \geq \frac{J-1}{L_2 - J + 1}(L-1).$$

For the case that there are as many microphones as there are sources ($L_2 = J$) separation is possible. The deconvolution filters must have $J-1$ times as many coefficients as the mixing filters in this case however. This can put severe demands on the separation algorithm for increasing J.

Deconvolution

Deconvolution includes both signal separation and dereverberation. For perfect deconvolution, it is required for the stacked convolution matrix to be full rank. It is assumed that the length of the acoustic paths L is estimated correctly and that the minimum delay in the direct acoustic paths is accounted for to prevent the transfer matrix from becoming singular, see Nelson et al., 1995. From a practical point of view, this means that the desired response must be time aligned so that it covers the same time-span as the cascaded mixing and unmixing process.

The transfer matrix reduces in rank by 1 when all acoustic transfer functions from a source to all microphones share a common zero. This can be seen by investigating the existence of a column that is a linear combination of all other columns of (2.2). This can be written as

$$\sum_j \sum_l \alpha_{jl} z^{1-l} h'_{ji}(z)(1 - n_c z^{-1}) = 0 \tag{2.4}$$

with $h'_{ji}(z)(1 - n_c z^{-1})$ the z-transform of $\underline{h}^L_{ji}$ and α_{jl} the linear combination coefficients. The common zero is expressed by $1 - n_c z^{-1}$ and the remainders of the transfer functions are in $h'_{ji}(z)$. Without solving this equation it is readily clear that a zero that is shared among all transfer functions from the ith source to all microphones plays no part in the rank of the transfer matrix as it can be divided out of (2.4). Note that this procedure can be repeated for all common zeros.

When only separation is considered, it is not a problem if the reduced stacked convolution matrix is not full rank. To achieve separation only, the filter vectors that are sought are chosen such that they are in the null space of the reduced stacked convolution matrix. This is due to the desired impulse responses from s_i to y_j being zero $\forall i \neq j$. When the null space becomes larger, the problem is relaxed even further. When the rank is reduced by r, the null-space of this matrix of this matrix increases accordingly. Therefore blind signal separation of signals that are mixed by a mixing system that is rank deficient can be done with a filter length of

$$N \geq \frac{(J-1)(L-1) - r}{L_2 - J + 1}.$$

When the stacked convolution matrix is not full-rank, joint dereverberation and separation is not possible however so that the source signals cannot be recovered exactly.

Some experiments on the trade off between the number of microphones used and the number of filter coefficients that is needed to obtain a good inverse are reported in Westner, 1998.

2.5. Noise Cancellation

In sound reproduction systems, sound signals are filtered prior to reproduction by loudspeakers to ideally obtain perfect sound reproduction at a finite number of positions in space. The filters can be found by first placing microphones at the relevant positions and using the difference between the ideal sound and the reproduced sounds at these positions as errors signals for an adaptive algorithm. The difference with classic adaptive filter theory is that these error signals are obtained by comparing the desired signals with adaptive filter outputs that are filtered by acoustic transfer functions instead of with the adaptive filter outputs themselves. The classical adaptive filter is depicted in Figure 2.6 (*top*) where it is used to equalize the acoustical transfer function $H(z)$. Here, the update uses the input signal of the adaptive filter $W(z)$ and the difference between the reference signal $d[n]$ which resembles $x[n]$ and the adaptive filter output. Instead of equalizing a signal $x[n]$ that is attenuated by an acoustic transfer function $H(z)$ sound reproduction systems need to filter this signal prior to transmission as depicted in Figure 2.6 (*bottom*). Both systems are equivalent if the adaptive filter $W(z)$ is constant. In practical applications it suffices to demand that the adaptive filter is slowly varying. The latter system is termed

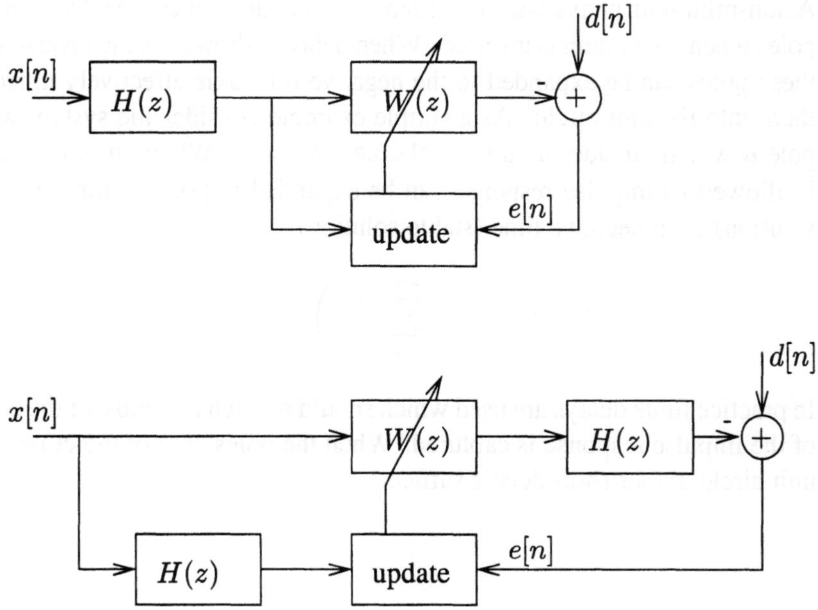

Figure 2.6. Conventional adaptive filter (*top*) and filtered-x equivalence (*bottom*)

the filtered-x in Nelson and Elliott, 1992 which indicates that a filtered version of the signal $x[n]$ is used in the update. This filter $H(z)$ which corresponds to the acoustic transfer function is unknown in practical situations. An estimate of it can be used however, and the filtered-x algorithm is known to be robust to estimation errors herein. Application areas include noise cancellation in ducts and phantom source generation Garas, 1996.

Notes

1 Setting the derivative of the cost function to zero yields the optimal weight vector

$$\underline{w}_{\mathrm{opt}}^{JN} = -(R_x^{JN})^{-1} C^{JN,p} \underline{\Lambda}^p$$

An expression for the Lagrange multipliers $\underline{\Lambda}^p$ is found by demanding that this optimal weight vector satisfies the constraints

$$(C^{JN,p})^T \underline{w}_{\mathrm{opt}}^{JN} = \underline{f}^p \,.$$

2 A non-minimum phase system has zeros outside the unit circle which become poles when the system is inverted. When delay is allowed in the inverse filter, these poles can be expanded to the negative time axis effectively mirroring them into the unit circle. As a simple example consider the system with 1 pole a which outside the unit circle, i.e. $|a| > 1$. When an infinite delay is allowed its impulse response can be expanded in positive time (unstable solution) or in negative time (stable solution)

$$\sum_{n=1}^{\infty} a^n = \sum_{n=-1}^{-\infty} \left(\frac{1}{a}\right)^n$$

In practice finite delays are used which should be such that most of the energy of the impulse response is captured. When the poles are not too close to the unit circle, rather short delays suffice.

Chapter 3

EFFICIENT FILTERING USING FFTS

This chapter gives an introduction to the efficient implementation of filters in the frequency domain. A background of design aspects is given to facilitate for the real-time implementation of such filters for a given latency (i.e. the delay between the input and the output of a filter). This chapter may be skipped when the reader is familiar with the overlap-save technique in the frequency domain and has no interest in new insights on low latency frequency domain filters.

Large (adaptive) filters can be implemented efficiently in the frequency domain using FFTs. Adaptive filters contain a learning part which updates the filter coefficients based on a reference signal or target function. These coefficients are used by the filter part to produce the output signal of the adaptive filter. The latency of the frequency domain implementation is typically larger that that of the time domain implementation. The design objectives of the filter part typically are to have a low number of operations per second, a low latency and a low memory usage.

The algorithms that are presented in the next chapters have in common that they employ Finite Impulse Response (FIR) filters. These filters are implemented in the frequency domain to enable real-time operation. In general, this fast implementation is applicable to all algorithms that employ FIR filters. FFTs based filters are known for a long time and are treated in text books, e.g. see Haykin, 1996. For a more detailed treatment of FFT based adaptive filters see Sommen, 1992. Information on processing and algorithm delay of various implementations and their corresponding complexity is crucial when choosing an implementation. When long FIR filters are used to enhance audio for television sets for example, straightforward FFTs implementations may result in a noticeable delay between audio and video. In some active noise cancellation applications, the filter needs to have a very low latency in order to create anti-noise that is time aligned with the physical noise signal.

The FIR filter is a linear convolution, whereas the update rule of the learning part of an adaptive filter is a linear correlation. This chapter is concerned only with linear convolutions, but the efficient implementation of linear correlations is likewise, as will be explained in the next chapter. Expressions for algorithm and processing delays are given next and it is explained how latency can be traded for computational complexity.

3.1. FIR Filters

In some applications adaptive filters are used to identify an unknown system that contains few parameters and has an infinite length impulse response. Adaptive filters that use infinite impulse response filters are reported in literate for this purpose, e.g. see Belt et al., 1995. When the filter structure of the adaptive filter accurately matches that of the unknown system, system identification can become much more fast and accurate. Acoustic transfer functions cannot be estimated accurately using only few parameters however due to their inherent complexity. This makes the FIR filter structure a more attractive candidate for an adaptive filter in acoustic applications.

Latency and computational complexity are given next for (multichannel) FIR filters that are implemented in the time domain. A FIR filter is defined as

$$y[n] = \sum_{l=0}^{N-1} w[l]x[n-l]\,. \tag{3.1}$$

It is stable by definition given that the energy of its coefficients is finite, i.e.

$$\sum_{l=0}^{N-1} (w[l])^2 < \infty$$

A straightforward time domain implementation requires N multiplications and N additions per computed output sample. Most processors typically make use of a combined multiply-add instruction when implementing time domain filters, so that a good measure of the number of instructions per second that is required can be the number of multiplications per second

$$\Psi_{\text{FIR}}\{N, f_s\} = N f_s\,,$$

with f_s the sample frequency. As an example, $N = 4096$ and $f_s = 44100$ Hz so that $\Psi_{\text{FIR}}\{4096, 44100\} = 180$ MIPS (Million Instructions Per Second). The filter length that is required in an adaptive filter is related to the reverberation time of a room in acoustic applications. With a length of 4096 filter tabs at fs=44100 Hz the FIR filter can capture 4096/44100 $\approx$ 0.1 seconds of an acoustic impulse response which can be rather short for many applications/rooms. From this it is clear that real-time implementation of many of such systems is too expensive for most consumer applications today. For portable applications, power consumption specifications are also typically not met. These problems grow even larger for multichannel systems. A multichannel FIR filter which has FIR filters between each of its L_1 inputs and L_2 outputs is defined as

$$y_k[n] = \sum_{l=0}^{N-1} \sum_{m=1}^{L_1} w_{km}[l] x_m[n-l] \text{ for } k = 1 \ldots L_2$$

and requires $L_1 \times L_2$ the complexity of a single channel FIR filter

$$\Psi_{\text{MC-FIR}}\{N, L_1, L_2, f_s\} = L_1 L_2 \Psi_{\text{FIR}}\{N, f_s\}.$$

The cost of such a filter is quite high and real-time implementation can be too expensive for most applications.

The time-domain implementation of the FIR filter has the advantage that it has an algorithm delay of only 1 sample. In general, the total delay between the input and output signal of a filter implementation consists of

- Algorithm delay. This delay is inherent to the filter algorithm that is used and it is always present, no matter how fast the processing core is, no matter how low the group delay of the filter is or the A/D D/A delay is. In (3.1) the algorithm delay theoretically equals 0 samples. It is more practical to consider this algorithm delay to be equal to 1 sample however, as it is not possible to output the result the very same sample as the input sample was obtained. The algorithms of the next sections work on blocks of data, so that the processor needs to wait for a block of data to be inputted before it can compute the output signal. The block size can be several thousands of taps in an efficient implementation of an FIR filter, typically resulting in an algorithm delay of about 0.1 seconds (e.g. 4096 taps at a sampling frequency of $f_s = 44100$ Hz). In audio-visual systems (e.g. teleconferencing) the audio should not be delayed by more than about 20 ms with respect to the video stream as this will result in the loss of lip synchronization.

- Processing delay. This is the time that the processor takes to compute the output signal. The processors capabilities are best exploited when the load is constant. When computing the filter in real-time is the only task of the processor, the processing delay is typically chosen to be equal to the algorithm delay. This is illustrated in Figure 3.1 where the processing delay is smaller that the algorithm delay resulting in a high peak load. A more economic implementation would have a constant load to which the processor's performance is matched. The processing of a block of data can therefore best take the same amount of time as the processor needs to wait for new data coming in, i.e. the algorithm delay.

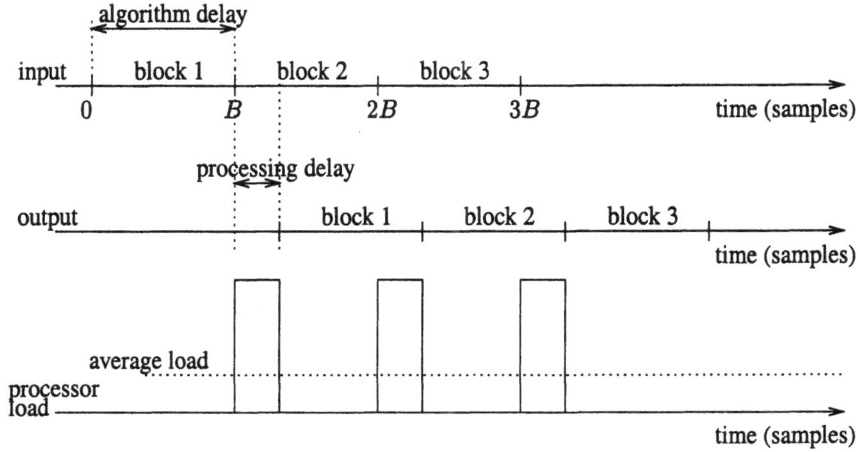

Figure 3.1. Relation between processing delay and processor load

- Group delay. The filter that is implemented has a group delay which can be anywhere between 0 and the filter length. When the filter equals the unit impulse for example, the group delay of the filter is equal to the position of this unit impulse.

- A/D (if equipped) and D/A delay. Most systems contain A/D and D/A converters to go from analog continues time and amplitude signals to discrete time and amplitude signals and visa versa. These A/D and D/A converters contain low pass filters to prevent aliasing. Some systems are critically sampled in the sense the signals' bandwidth is close to the Nyquist frequency (i.e. half of the sampling frequency). The low pass filters need to have a sharp transition in that case, resulting in relatively large group delays of about 1 ms.

3.2. Block Based FIR Filters

Finite impulse response filters can be implemented to work on blocks of data at a time, readily improving parallelism when implemented in the time domain. Effectively the input signal is segmented into blocks which can be processed independently. The blocks are processed (convolved) yielding blocks of output data. These blocks are joined together in a proper way to obtain the output signal. Overlap-add and overlap-save are two well known methods for doing so, see Ferrara, 1980. They both have about the same computational complexity, although overlap-save has somewhat more flexibility in choosing its parameters, see Sommen and Jayasinghe, 1988. Techniques in this book are based on overlap-save.

Overlap-add

Overlap-add is explained next for fixed filters. When two vectors $\underline{X}^M$ and $\underline{W}^M$ are convolved in the frequency domain as

$$\underline{y}^M = \mathcal{F}^{-1}(\underline{X}^M \otimes \underline{W}^M),$$

the length $2M-1$ convolution result is folded back into a length M vector y^M, effectively destroying all samples but one. Overlap-add uses zero padding of both the data and the filter. In the example that is shown in Figure 3.2, a filter w of length $N = 4$ is augmented with $M - N$ zeros so that the convolution can be carried out using (I)FFTs of length $M = 8$. Note that these short filter and FFT sizes are not practical, they are used only to explain the techniques. A segment of $B = M - N + 1 = 5$ samples is taken from an infinite length

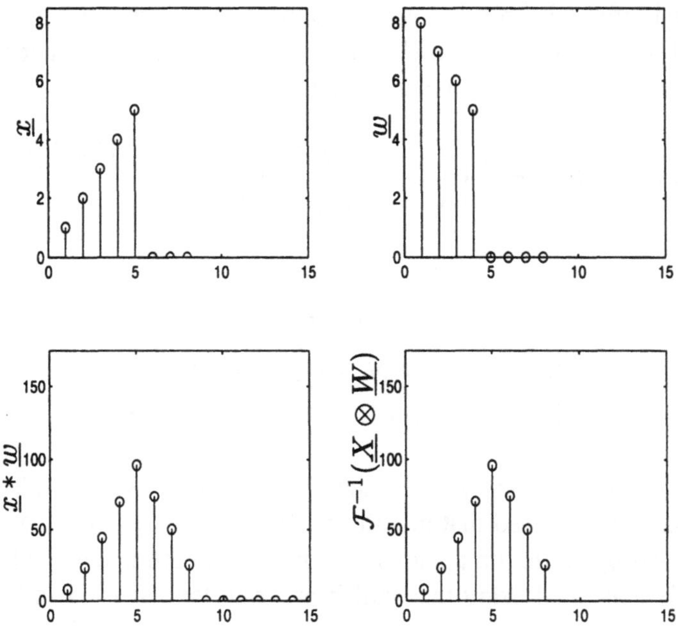

Figure 3.2. Example illustrating the overlap-add method for one data segment

data series which is augmented with 3 zeros to length M. As a reference, the linear convolution of the augmented filter and data $\underline{x} * \underline{w}$ is shown in the lower left of the figure. Although this convolution result is of length $2M - 1$, only $N + B - 1$ samples are relevant, the others equal zero due to the augmentation with zeros. When this convolution is computed in the frequency domain using length 8 (I)FFTs as is shown in the lower right part of the figure, the $M - 1$ samples in the convolution result that are folded back are equal to zero and do therefore not alter the relevant samples.

```
x = sin(1:2*pi/9:30)';         % input signal
y = zeros(size(x));
N = 10;                        % filter length
M = 20;                        % FFT size
B = 11;                        % number of new samples per block (B=M-N+1)
w = ones(N,1);                 % the filter
imax = length(x)-M+1;          % max. iterations
W = fft([flipud(w);zeros(M-N,1)]);     % frequency domain filter
for i = 1:B:imax,
  X = fft([x(i:i+B-1);zeros(N-1,1)]); % frequency domain data
  y_temp = real(ifft(W.*X));          % circular convolution
  y(i:i+M-1) = y(i:i+M-1)+y_temp(1:M);
end
```

Figure 3.3. Matlab routine explaining overlap-add

To filter a infinite length data series, the next $B = 5$ samples are taken and augmented with zeros. The convolution is computed in the frequency domain and added to the previous convolution result

$$\underline{y}^M[kB] = \underline{y}^M[kB] + \mathcal{F}^{-1}\left(\left(\mathcal{F}\left(\begin{matrix}\underline{x}^B[kB]\\ \underline{0}^{N-1}\end{matrix}\right)\right)\otimes\left(\mathcal{F}\left(\begin{matrix}\mathrm{J}^N\underline{w}^N[kB]\\ \underline{0}^{B-1}\end{matrix}\right)\right)\right),$$

with

$$\begin{aligned}\underline{x}^B[kB] &= (x[kB-B+1]\ldots x[kB])^T\\ \underline{w}^N[kB] &= (w[N-1]\ldots w[0])^T\\ \underline{y}^M[kB] &= (y[kB-M+1]\ldots y[kB])^T\end{aligned}$$

and where the mirror matrix J^N flips the filter vector $\underline{w}^N[kB]$ upside down. In this equation it is assumed that initially $\underline{y}^M[kB] = \underline{0}^M$ for all k. An example of overlap-add operating on a signal x of infinite length is depicted in Figure 3.4 which is generated using the Matlab program that is listed in Figure 3.3. The figure shows three partial convolution results $\underline{y}_1$ to $\underline{y}_3$ and $\underline{y}$ which is the sum of these partial convolutions. As $\underline{y}_4$ is not computed yet in this Figure, $\underline{y}$ is identical to $\underline{x} * \underline{w}$ except for the last few samples. Note that the Matlab code assumes $B = M - N + 1$ as the relation between the number of new samples B, the filter length N and the Fourier transform length M. Less new samples B per block may also be used in general $B \leq M - N + 1$, requiring that some of the computed data should be thrown away.

When the filter is changing in time, the current filter should be convolved with both the current and the previous data segment in order to reconstruct B output samples that correspond with the current filter. This can still be done with a similar efficiency when $M = 2B = 2N$, see Sommen and Jayasinghe, 1988.

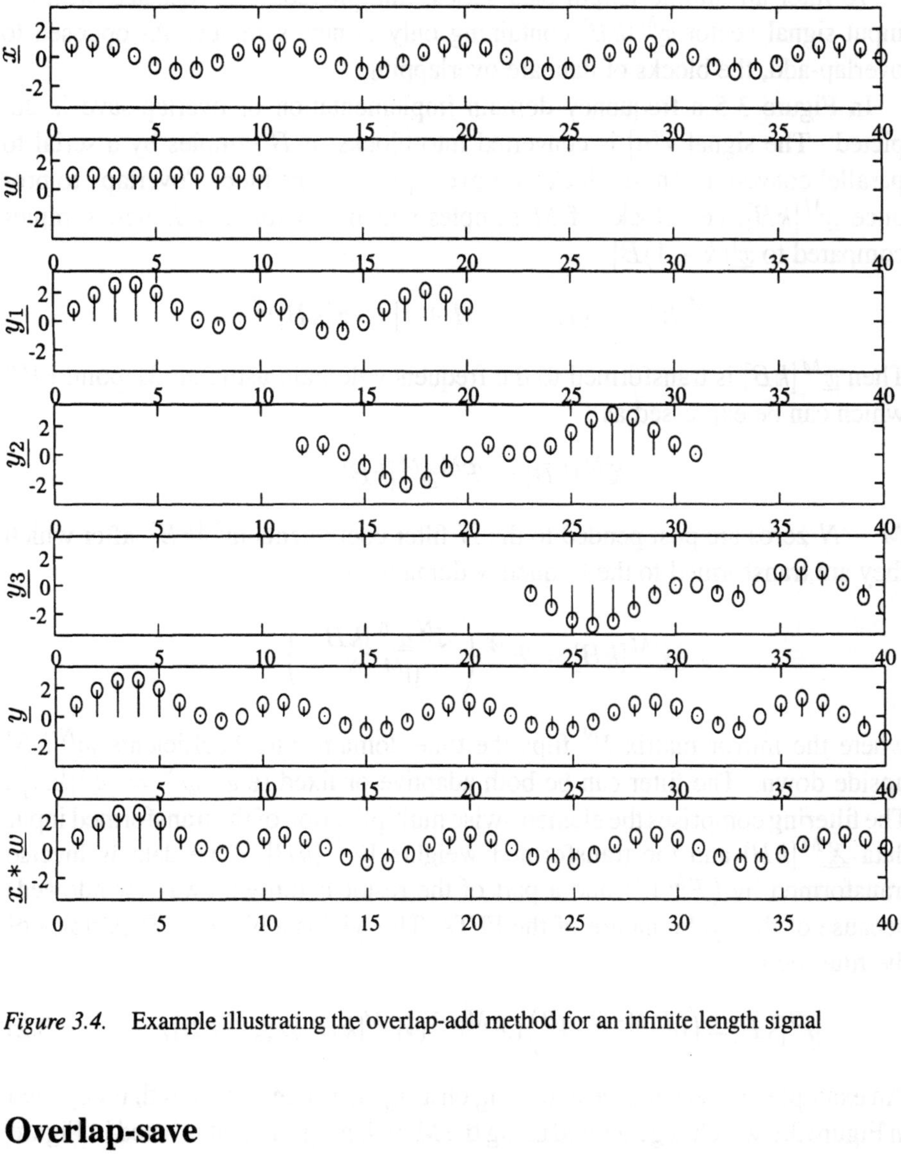

Figure 3.4. Example illustrating the overlap-add method for an infinite length signal

Overlap-save

In the overlap-save method the filter coefficients are augmented with zeros but the data is not. The circular convolution result will therefore wrap around and destroy some of the output samples. These samples are thrown away and the next block of input samples is chosen to be overlapping with the previous one in order to compensate for this "loss". In the example that is shown in Figure 3.8, a filter w of length 4 is augmented with $M - N$ zeros so that the convolution can be carried out using (I)FFTs of length $M = 8$. A segment of $M = 8$ samples is taken from an infinite length data series. When the circular convolution is compared to the linear one it can be seen that only sample 4 to 8 ($B = 5$ samples) can be used from the circular convolution result.

To filter an infinite length data series, this procedure is repeated with the input signal vector $\underline{x}^M[kB]$ containing only B new samples. As opposed to overlap-add, the blocks of data are overlapping.

In Figure 3.5 a frequency domain implementation of overlap-save is depicted. The signal $x[n]$ is converted into blocks of B samples by a serial to parallel converter. These blocks are overlapped by the block "overlap" to produce $\underline{x}^M[kB]$, i.e. blocks of M samples which contain only B new samples compared to $\underline{x}[(k-1)B]$

$$\underline{x}^M[kB] = (x[kB - M + 1] \dots x[kB])^T .$$

Then $\underline{x}^M[kB]$ is transformed to the frequency domain using an M point FFT, which can be expressed as

$$\underline{X}^M[kB] = \mathcal{F}^M \underline{x}^M[kB].$$

$M - N$ zeros are post pended to the N filter coefficients $\underline{w}^N[kB]$, after which they are transformed to the frequency domain, i.e.

$$\underline{W}^M[kB] = \mathcal{F}^M \begin{pmatrix} \mathrm{J}^N \underline{w}^N[kB] \\ \underline{0}^{M-N} \end{pmatrix}$$

where the mirror matrix J^N flips the time domain filter coefficients $\underline{w}^N[kB]$ upside down. The filter can be both adaptive or fixed (e.g. $\underline{w}^N = \underline{w}^N[kB]$). The filtering comprises the element-wise multiplication of the transformed input data $\underline{X}^M[kB]$ and the transformed weights $\underline{W}^M[kB]$. The data is inverse transformed by $(\mathcal{F}^M)^{-1}$ and a part of the result is thrown away (windowed) because of the cyclic nature of the FFTs. This yields a block of B samples of the filter output

$$\underline{y}^B[kB] = \begin{pmatrix} 0^{B,M-B} & \mathrm{I}^B \end{pmatrix} (\mathcal{F}^M)^{-1} (\underline{W}^M[kB] \otimes \underline{X}^M[kB]). \tag{3.2}$$

An example of overlap-save operating on a signal x of infinite length is depicted in Figure 3.9 which is generated using the Matlab program that is listed in Figure 3.7. The convolution result $\underline{y}$ is composed of the last B samples of $\underline{y}_1$, $\underline{y}_2$ and $\underline{y}_3$.

3.3. Low Delay Frequency Domain FIR Filters

The algorithm delay of overlap-save is B samples as follows from (3.2). The processing delay is typically also B samples unless the processor has many other tasks besides computing the filter and is able to distribute them well in time. The group delay depends entirely on the filter that is implemented ranging from 0 to the filter length N. Let's assume for example that the group delay equals $N/4$. It is assumed further that the A/D D/A delay equals 1 ms and

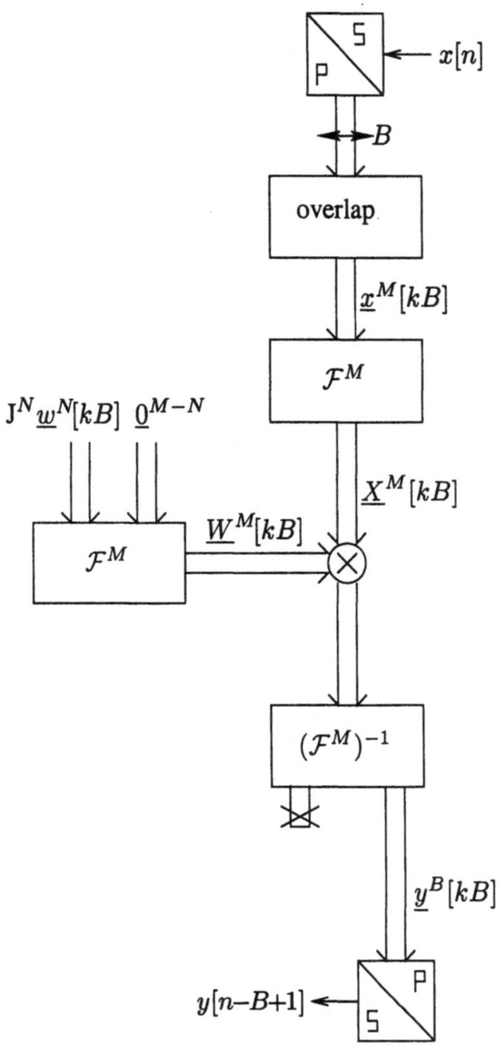

Figure 3.5. FIR filter implemented using the overlap-save method in the frequency domain

that the total delay should less that 20 ms for a given audio-video application. The computational complexity depends on the sampling frequency. The delay (except for the processing delay perhaps) does not depend on the sampling frequency as the delay and all its relevant parameters (filter length etc.) are expressed in time (milliseconds). It is convenient to introduce a certain sampling frequency however, so that a numerical example can be given for all parameters. In this example that the sampling frequency is $f_s = 44100$ Hz and the filter is of length $N = 2048$ (about 50 ms). The admissible delay that remains for algorithmic and computational delay is 20 ms (delay budget) - 11.6 ms (group delay of $N/(4fs)$) - 1 ms (A/D D/A delay) = 7.4 ms. This corresponds to 326

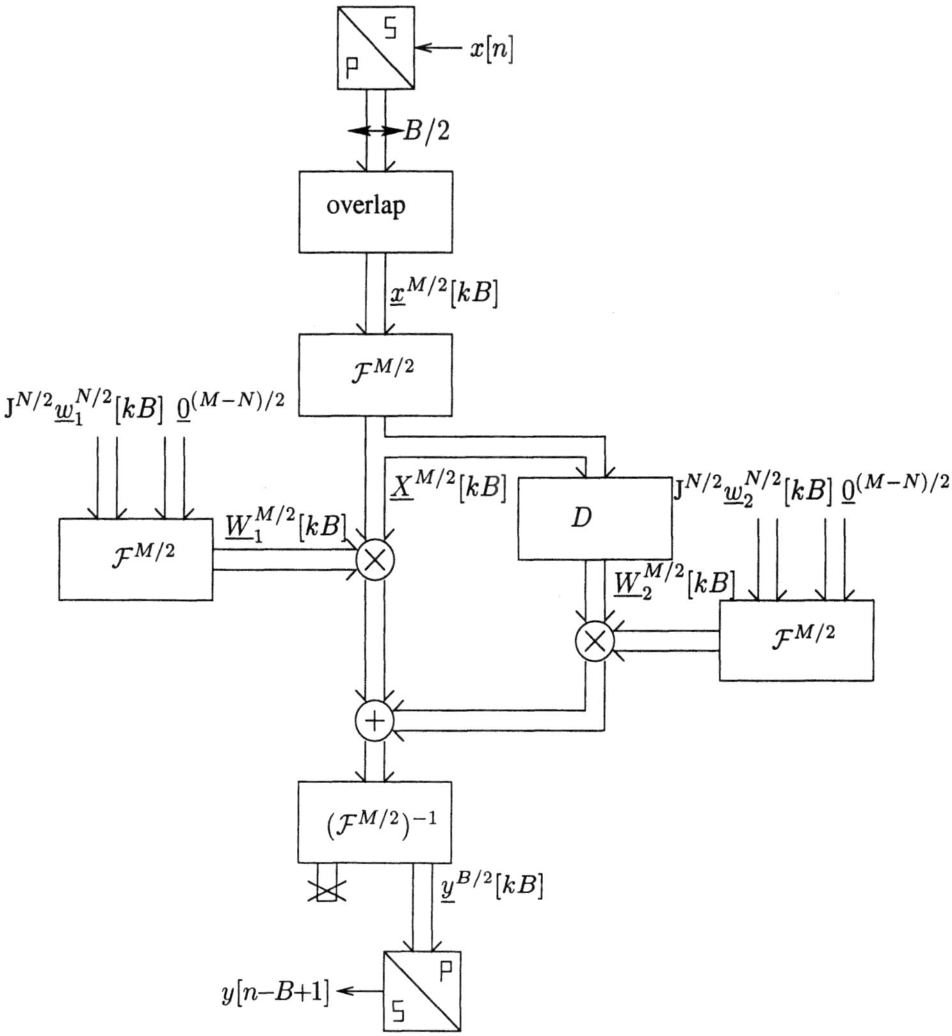

Figure 3.6. Segmented FIR filter implemented using the overlap-save method in the frequency domain

samples. In order for the (I)FFTs to be implemented efficiently their length is a power of 2. The (I)FFT length is therefore chosen to be $M = 4096$ in this case which is the smallest power of 2 that is larger than the filter length N. Every block $M - N + 1 = 2049$ new samples are computed. From these 2049 samples, only the 163 newest samples are used each block as the admissible delay delay for both computational delay (B samples) and algorithmic delay (B samples) is only 326 samples. Despite the fact that overlap-save computes 2049 samples each block, it can only use $B = 163$ new input samples per block as

```
x = sin(1:2*pi/9:30)';          % input signal
yos = zeros(size(x));
N = 10;                         % filter length
M = 20;                         % FFT size
B = 11;                         % number of new samples per block (B<=M-N+1)
w = ones(N,1);                  % the filter
imax = length(x)-M+1;           % max. iterations
W = fft([flipud(w);zeros(M-N,1)]); % frequency domain filter
for i = 1:B:imax,
  X = fft(x(i:i+M-1));                  % frequency domain data
  y_temp = real(ifft(W.*X));            % circular convolution
  y(i+M-B:i+M-1) = y_temp(M-B+1:M);
end
```

Figure 3.7. Matlab routine explaining overlap-save

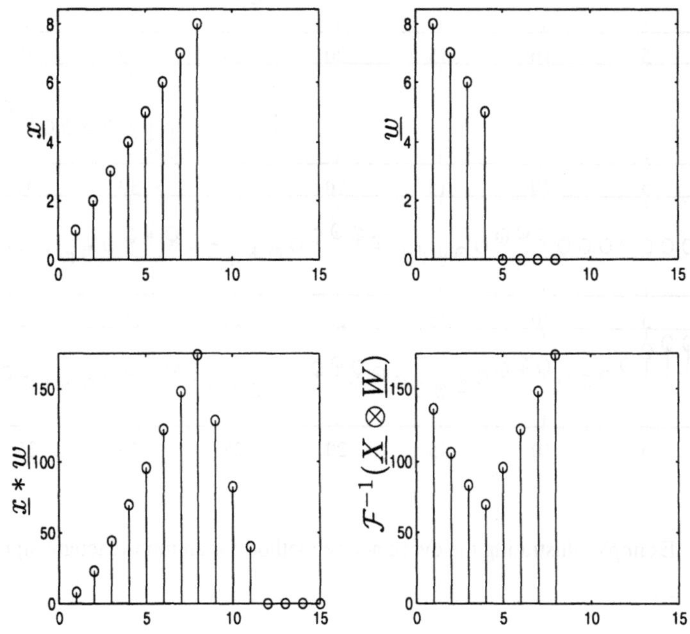

Figure 3.8. Example illustrating the overlap-save method for one data segment

only 163 output samples are used. It is clear that the resulting implementation is not very efficient.

Segmented overlap-save

The FIR filter is split into two parts in Figure 3.6. The block 'D' delays the frequency domain data $\underline{X}^{M/2}[kB]$ over $B/2$ samples. The algorithm delay is reduced from B in Figure 3.5 to $B/2$ in Figure 3.6. It is generally believed

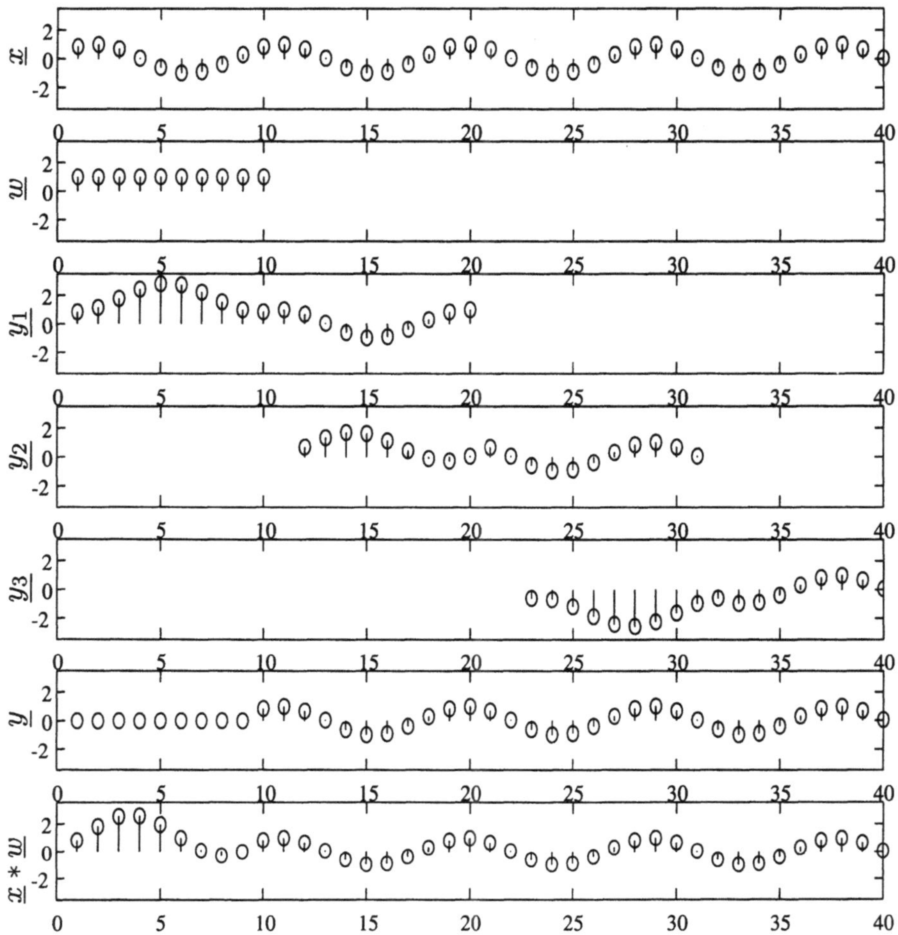

Figure 3.9. Example illustrating the overlap-save method for an infinite length signal

that reducing the latency in this way increases the computational complexity. It is shown next that in some cases the computational complexity is even lower when the filter is segmented.

In the segmented implementation, the (I)FFTs are of length $B/2$ which can be computed more efficiently than the length B (I)FFTs but they have to be computed twice as often. In Sorenson et al., 1987 quite some attention is paid to optimization details and a computational complexity of $M\log_2(M) - \frac{7}{2}M - 6$ multiplications and $\frac{3}{2}M\log_2(M) - M/2 - 2$ additions is reported for the implementation of a length M FFT. The sum of all additions and multiplications that are reported in Sorenson et al., 1987 will be used as a measure for compu-

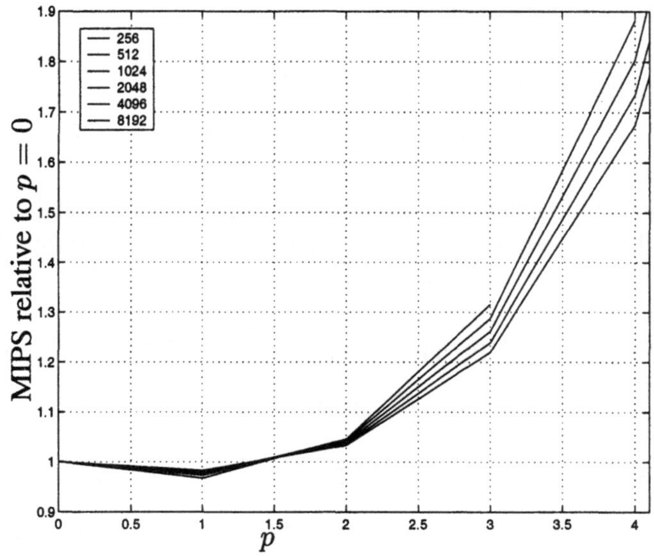

Figure 3.10. Relative MIPS loss when segmenting a filter in 2^p parts

tational complexity in the following derivations

$$\Psi_{\text{FFT}}\{M\} = \tfrac{5}{2}M\log_2(M) - \tfrac{9}{2}M - 8$$

The complexity per output sample of a filter that is implemented in the frequency domain using a length M (I)FFT is

$$\Psi_{\text{FILTER}}\{M\} = \tfrac{2}{M}\left(2\cdot\Psi_{\text{FFT}}\{M\} + \Psi_{\text{multiply}}\{M\}\right)$$

where it is assumed for simplicity that $\frac{M}{2}$ samples are computed in every block and where Ψ_{multiply} is the complexity of the complex element-wise multiplications (this involves $\frac{M}{2} - 1$ complex and 2 real multiplications)

$$\Psi_{\text{multiply}}\{M\} = 6(\tfrac{M}{2} - 1) + 2$$

Note that a complex multiplication can be done using 6 real operations. When the filter is segmented into 2^p pieces of equal length, the complexity becomes

$$\begin{aligned}\Psi_{\text{FILTER}}\{M,p\} = \tfrac{2\cdot 2^p}{M}\Big(&2\cdot\Psi_{\text{FFT}}\{\tfrac{M}{2^p}\} + 2^p\cdot\Psi_{\text{multiply}}\{\tfrac{M}{2^p}\} \\ &+(2^p-1)\cdot\Psi_{\text{additions}}\{\tfrac{M}{2^p}\}\Big) \qquad (3.3)\end{aligned}$$

with $\Psi_{\text{additions}}\{M\}$ the number of real additions that are required to add the results of the subfilters (this involves $\frac{M}{2} - 1$ complex and 2 real additions)

$$\Psi_{\text{additions}}\{M\} = M$$

Evaluating (3.3) further gives

$$\Psi_{\text{FILTER}}\{M,p\} = \tfrac{2\cdot 2^p}{M}\left(2\left(\tfrac{5}{2}\tfrac{M}{2^p}\log_2\{\tfrac{M}{2^p}\} - \tfrac{9}{2}\tfrac{M}{2^p} - 8\right)\right.$$
$$\left. +2^p\left(6(\tfrac{M}{2\cdot 2^p} - 1) + 2\right) + 2(2^p - 1)\tfrac{M}{2\cdot 2^p}\right)$$

An expression can be given now for the increase in computational complexity when a filter that is segmented into 2^p parts will be segmented further into 2^{p+1} parts

$$\Psi_{\text{FILTER}}\{M,p+1\} - \Psi_{\text{FILTER}}\{M,p\} = -10 + 2^p(8 - \tfrac{32}{M}) - 2^{2p}\tfrac{24}{M}$$

Evaluating this expression for $p = 0$ gives

$$\Psi_{\text{FILTER}}\{M,1\} - \Psi_{\text{FILTER}}\{M,0\} = -2 - \tfrac{56}{M}$$

This increase in complexity is always negative, for all M, meaning that in fact computational complexity is saved when the filter is segmented into two pieces. For $p = 1$ the increase in complexity is

$$\Psi_{\text{FILTER}}\{M,1\} - \Psi_{\text{FILTER}}\{M,0\} = 6 - \tfrac{160}{M}$$

which is always positive (but small) for practical values of M. For larger values of p the computational complexity increase relative to $\Psi_{\text{FILTER}}\{M,0\}$ can be seen from Figure 3.10 for $M = 256, 512, 1024, 2048, 4096, 8192$. When $p = 2$ for example, the algorithm and processing delay are 4 times as low compared to $p = 0$ (no segmentation), at the cost of a computational complexity that is approximately 5% higher. Note that p should not exceed $\frac{\log_2\{M\}}{\log_2\{M_s\}}$, with M_s the smallest economic FFT size (e.g. $M_s = 32$).

An excellent paper on the topic of FIR filter segmentation is given in Gardner, 1995, where it is assumed that a larger FFT size gives rise to a lower computational complexity. An economic segmentation is therefore

$$[N; N; 2N; 4N; 8N; 16N; \ldots].$$

It is shown in this paper that this segmentation is not practical however in terms of processing delay. A more practical segmentation is shown to be

$$[N; N; 2N; 2N; 4N; 4N; \ldots].$$

which has a slightly higher computational complexity than the first one. Also, it is exploited that a length $2N$ FFT can be computed efficiently when length N FFTs have already been computed for the same data.

It is clear however from Figure 3.10 that reusing the same FFT size many times can be very beneficial. It is shown next that a segmentation with relatively few different FFT sizes is therefore much more economic. An optimal segmentation can be found using a computer program that evaluates all possible scenarios. In order to obtain more insight in the segmentation and to find a practical solution close to the optimum solution, a greedy optimization is proposed next. The greedy filter optimization algorithm is shown in Figure 3.11. It starts with a filter of length $N = 128$ which corresponds to a algorithm delay of about 3 ms. The filter can be made longer without increasing the algorithm delay by adding a subfilter of the same length as in Figure 3.6. The additional computational complexity is only due to the additional element-wise multiplication and the additions as the (I)FFTs can be shared. The filter can be extended with more and more subfilters of length 128, but at some point it will be more efficient to use a larger (I)FFT size for the next blocks. One way to decide whether a larger FFT size is cheaper than to hold on to the previous FFT size is to compare the computational complexity per computed output sample using the larger FFT size to the computational complexity per output sample of the element-wise multiplication and the additions that are required to reuse the previous (I)FFTs. This is what the greedy algorithm does when 'ratio'=1. Once a larger FFT size is used, the next blocks will also benefit from this. Therefore the greedy algorithm should be stimulated to choose a larger FFT size sooner by setting 'ratio'=0.5 for example. The influence of 'ratio' can be seen from Table 3.1. In the most efficient segmentation of Table 3.1, the length 8192

Table 3.1. Number of subfilters and MIPS of the greedy optimization algorithm with a total filter length $N = 8192$ and $fs = 44100$ for various ratios.

'ratio'	#subfilters	MIPS
0-0.1	7	28
0.2	9	18
0.3	12	15
0.4-0.6	15	12
0.7-1.1	17	14
1.2	33	19

filter is split in 15 subfilters:

[128;128;128;128;128;128;128;128;1024;1024;1024;1024;1024;1024;1024]

Alternatively, if every FFT size is used twice as in Gardner, 1995, a length 7936 filter can be made with the following segmentation

$$[128; 128; 256; 256; 512; 512; 1024; 1024; 2048; 2048]$$

The scheduling will obviously be more complicated and the complexity is 18 MIPS which is about 50 % more than the segmentation that is found by the greedy optimization algorithm.

```
fs = 44100;
N = [128];    % smallest subfilter size
ratio = 0.5; % used to decide to increase FFT size or not

% complexity per sample for a length N filter:
phi = [2/N*(5/2*2*N*log2(2*N)-9/2*2*N-8) + 1/N*((N-1)*6+2)];

for count=1:64,
  % keep same FFT size:
  Na = N(length(N));
  % complexity per sample for a length Na filter:
  phi_a = 1/Na*((Na-1)*6+2) + 2;

  % use largest possible FFT size:
  Nb = 2^floor(log2(sum(N)));
  % complexity per sample for a length Nb filter:
  phi_b = 2/Nb*(5/2*2*Nb*log2(2*Nb)-9/2*2*Nb-8) + 1/Nb*((Nb-1)*6+2) + 1;

  if (phi_a/Na < ratio*phi_b/Nb ) % use larger FFT?
    phi = [phi;phi_a];
    N = [N;Na];
  else
    phi = [phi;phi_b];
    N = [N;Nb];
  end
end

sum(N(1:15))              % total filter length
fs*sum(phi(1:15))         % total complexity
```

Figure 3.11. Matlab program for greedy filter structure optimization

The results of the derivation that is presented in this section obviously depend on the assumptions that are made on the computational complexity. The general idea does not however; in a segmented filter the delay is reduced and (expensive) (I)FFTs are shared. Some additional computations are required to add the results of the subfilters together. It turns out to be favorable to reuse the same FFTs often rather than to increase the FFT size for every filter segment.

3.4. Conclusions

In this chapter frequency domain implementations of the overlap-add and overlap-save techniques are explained. Expressions for the latency are derived and methods for latency reduction are presented. In particular, it is shown that the latency can be halved without increasing the computational complexity and a corresponding practical filter segmentation is presented.

II

ACOUSTIC ECHO CANCELLATION

Chapter 4

AN EFFICIENT ADAPTIVE FILTER IMPLEMENTATION

Adaptive filters are used in many applications of which the most pronounced is probably echo cancellation. One example of such a filter is the LMS adaptive filter Sondhi, 1967; Widrow and Stearns, 1985. The LMS adaptive filter has always been quite popular because of its conceptual simplicity. The convergence can be extremely slow however when working with real-life audio signals and long filters. Moreover, the computational complexity grows linearly with the filter length making it impractical to implement long (thousands of coefficients) filters in real-time.

More recently, the Block Frequency Domain Adaptive Filter (BFDAF) is often used because of its superior performance when working with non-white input signals and its relatively low computational complexity.

The multichannel algorithms that are proposed in the following chapters use the same building blocks as the BFDAF so that the results that are obtained in this chapter also apply to these multichannel algorithms.

4.1. An Introduction to Adaptive Filters

An optimal filter can be designed to process stationary data based on statistical properties of the data that are known a priori. This filter is termed the Wiener filter after its inventor. In practical situations, this a priori information is often not fully present and the data to be processed is typically non-stationary. Adaptive filters can deal with such practical situations however as they are self-designing. In a stationary environment an adaptive filter converges, in the average, towards the optimal Wiener solution. When the environment is non-stationary it can track these changes and provides a time varying solution that will be rather good when the changes in the statistics of the signals are rather slow.

An adaptive filter is depicted in Figure 4.1. The known input signal x is filtered by an unknown system h producing a signal e. The observable signal $\breve{e}$

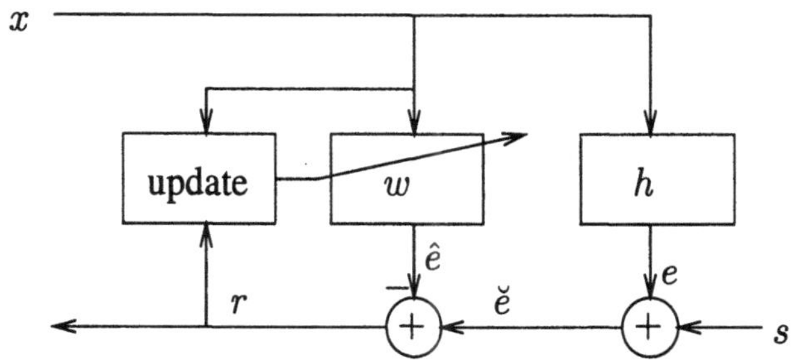

Figure 4.1. Signal estimation and system identification

equals the signal e together with a signal s which is assumed to be uncorrelated with x. The adaptive filter consists of the filter w that produces $\hat{e}$ which is an estimate of the observed signal $\breve{e}$. It further consists of an update part which recursively computes (updates) the filter coefficients w based on the input signal x and the residual signal r. The goal of the update is to minimize the residual signal r thereby effectively decorrelating the residual signal r with the input signal x. Assuming the the filter structure w is suitable to model the unknown system h this decorrelation will optimally lead to $w = h$ and $r = s$. Applications therefore obviously include system identification (finding h) and signal estimation (finding s). In Figure 4.2 the adaptive filter is depicted

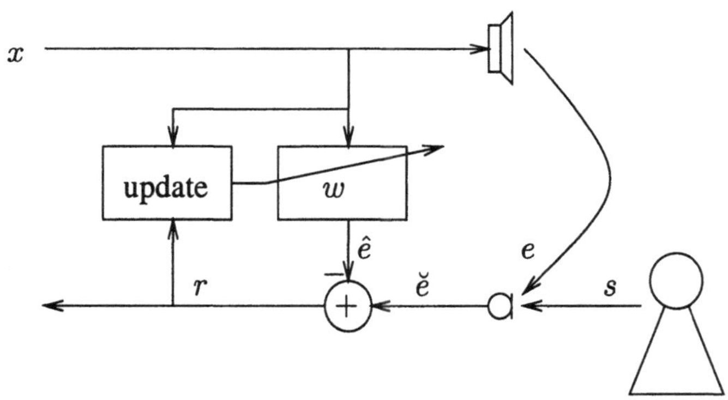

Figure 4.2. Acoustic echo canceler

for the application of acoustic echo cancellation. This is a signal estimation application where the voice of a speaker s is picked up by a microphone together with an acoustic echo from the loudspeaker that reproduces x. It is applied in mobile phones for example where x is the voice from the remote speaker that is amplified and reproduced through a loudspeaker. Echo cancellation is necessary to suppress the contribution of this loudspeaker signal to the microphone as was explained in South et al., 1979. Without echo cancellation the echo e is sent to the far-end signal where it will be amplified and sent back again to the near-end leading to an unacceptable round-singing. This problem is even worse in hands-free communication where the loudspeaker volume is much higher as the user is further away.

Other applications of echo cancellation include the data echo cancellation, signal correction and signal prediction. Data echo cancellation resembles acoustical echo cancellation and is used in modems. In signal correction an unknown desired signal is distorted by an unknown channel and contaminated with unknown noise. When there is a training sequence at regular times for example, an adaptive filter can try to restore the signal using the known training sequence as a reference signal. In signal prediction the adaptive filter makes an estimate of future samples of the signal based on previous ones. This technique can be used in signal compression for example, where only the unpredictable part needs to be transmitted or stored.

When designing an acoustical echo canceler four major problems arise. The first is that the acoustical echo path is long, typically 100-200 ms. When the sample frequency is 16 kHz or higher for example, the adaptive filter should have thousands of coefficients in order to be able to match the echo path well. Secondly, the echo path is typically time-varying. In a mobile phone application for example the user can change the position of the mobile phone slightly with respect to his/her head resulting in large changes in the echo path. Thirdly, the signal x is typically non-stationary. Finally, the adaptive filter is tuned to achieve fast convergence so that the signal s disturbs the performance of the adaptive filter significantly and can even lead to instability. Therefore, adaptation is typically stopped when there is a near-end speech signal s which is referred to as a double talk situation.

4.2. Time Domain Adaptive Filters

In this section background material is presented on time domain adaptive filters. The BFDAF is introduced by starting from this material and moving to the frequency domain. The near-end signal $s[n]$ is assumed to equal zero in the derivations that follow.

The update equation fully describes the adaptive filter and is usually obtained by minimizing the mean-squared error of the residual signal r

$$E\{r^2[n]\} = E\{((\underline{x}^N[n])^T(\underline{w}^N_{\text{opt}} - \underline{w}^N[n]))^2\}$$

with $\underline{w}^N_{\text{opt}}$ the Wiener solution for a length N adaptive filter. In this equation it is assumed that the unknown system is entirely represented by the length N Wiener solution so that $e[n] = (\underline{x}^N[n])^T\underline{w}^N_{\text{opt}}$. The derivative of the mean-squared error with respect to the adaptive weights is

$$\underline{\nabla}^N[n] = -2E\{\underline{x}^N[n]r[n]\}\,.$$

The classical LMS method is based on the steepest-descent method which updates the adaptive filter coefficients towards $\underline{w}^N_{\text{opt}}$

$$\underline{w}^N[n+1] = \underline{w}^N[n] - \mu\underline{\nabla}^N[n]$$

with μ a positive adaptation parameter. Approximating the expectation operation by the instantaneously value in the gradient yields the LMS algorithm

$$\underline{w}^N[n+1] = \underline{w}^N[n] + 2\mu\underline{x}^N[n]r[n]\,.$$

When the adaptation constant μ is small the filter converges slowly but does not fluctuate much when it arrives at the Wiener solution. A larger μ improves the convergence speed but has a consequence that the adaptive filter keeps jumping about the Wiener solution. When μ is too large, the adaptive filter becomes instable. The update speed depends on the power of the input signal

$$\sigma_x^2 = E\{(\underline{x}^N[n])^T\underline{x}^N[n]\}$$

as

$$E\{\underline{x}^N[n]r[n]\} = E\{\underline{x}^N[n](\underline{x}^N[n])^T(\underline{w}^N_{\text{opt}} - \underline{w}^N[n])\}\,. \tag{4.1}$$

This dependency can be removed by dividing the adaptation constant μ by an estimate of the power of the input signal $x[n]$ yielding the Normalized Least Mean Square (NLMS) algorithm

$$\underline{w}^N[n+1] = \underline{w}^N[n] + 2\frac{\mu}{\hat{\sigma}_x^2}\underline{x}^N[n]r[n]$$

Although the performance of NLMS is independent of the power of the input signal, it does not perform well if the input signal is correlated in time. In fact the normalization that is required to make (4.1) independent of the input signal is the inverse of the autocorrelation matrix of the input signal

$$E\{\underline{x}^N[n](\underline{x}^N[n])^T\}$$

In the Recursive Least Squares (RLS) adaptive filter the update is pre-multiplied by the inverse of an estimate of this autocorrelation matrix

$$\underline{w}^N[n+1] \;=\; \underline{w}^N[n] + 2\mu(\hat{R}_x^N)^{-1}\underline{x}^N[n]r[n]\,, \tag{4.2}$$

with $\hat{R}_x^N$ an estimate of the autocorrelation matrix. Although the performance of RLS is quite good, real-time operation is not feasible for large filters due to the matrix inversion. The decorrelation in RLS is approximately equivalent to normalizing each frequency component of the input signal x by an estimate of its power. The BFDAF is based on this observation and is an approximated RLS.

4.3. Block Frequency Domain Adaptive Filter

Large adaptive filters can be implemented efficiently in the frequency domain using FFTs, see for example Sommen, 1992. The BFDAF is such a frequency domain adaptive filter and is depicted in Figure 4.3. First the input signal $x[n]$ is segmented into blocks of length B by a serial to parallel converter. Next, the segments are overlapped

$$\underline{x}^M[kB] = (x[kB - M + 1] \ldots x[kB])^T .$$

The overlapping data is transformed to the frequency domain using

$$\underline{X}^M[kB] = \mathcal{F}^M \underline{x}^M[kB]. \tag{4.3}$$

The filtering comprises the element-wise multiplication of the transformed input data $\underline{X}^M[kB]$ and the transformed adaptive filter coefficients $\underline{W}^M[kB]$. The data is inverse transformed by $(\mathcal{F}^M)^{-1}$ and a part of the result is thrown away (windowed) because of the cyclic nature of the FFTs. This yields a block of B samples of the adaptive filter output

$$\underline{\hat{e}}^B[kB] = \left(0^{B,M-B}\ \mathrm{I}^B\right)(\mathcal{F}^M)^{-1}(\underline{W}^M[kB] \otimes \underline{X}^M[kB]), \tag{4.4}$$

with

$$\underline{W}^M[kB] = \mathcal{F}^M \begin{pmatrix} \mathrm{J}^N \underline{w}^N[kB] \\ 0^{M-N} \end{pmatrix} \tag{4.5}$$

and with J^N the $N \times N$ mirror matrix and $\underline{w}^N[kB]$ the time domain filter coefficients. To obtain linear convolutions, the size of the Fourier transform M must be equal to or larger than $N + B - 1$, with N the number of adaptive filter coefficients. The algorithm delay of the adaptive filter equals $B - 1$.

An efficient update is obtained by transforming the BRLS update of the previous section to the frequency domain. In the Block Frequency Domain Adaptive Filter (BFDAF) the energy of the input signal is normalized independently for all frequencies which increases the convergence speed drastically. Also, the computational complexity of both the filtering and the update is reduced significantly.

In order to move to the frequency domain where the convolution and correlation operations always work efficiently on blocks of data, the block RLS update is introduced

$$\underline{w}^N[(k+1)B] \;=\; \underline{w}^N[kB] + 2\mu(\hat{R}_x^N[kB])^{-1}\mathcal{X}^{N,B}[kB]\underline{r}^B[kB], \tag{4.6}$$

with $\hat{R}_x^N[kB]$ an estimate of the autocorrelation matrix that is given by

$$R_x^N \;=\; \frac{1}{B} E\{\mathcal{X}^{N,B}[kB](\mathcal{X}^{N,B}[kB])^T\}$$

$$\mathcal{X}^{N,B}[kB] = \begin{pmatrix} x[kB-B-N+2] & \dots & x[kB-N+1] \\ \vdots & & \vdots \\ x[kB-B+1] & \dots & x[kB] \end{pmatrix},$$

with $\underline{r}^B[kB] = (r[kB-N+1]\dots r[kB])^T$. The updates that would have been done in RLS update (4.2) are now accumulated first over B samples and than executed in one update.

In Egelmeers, 1995 it is argued that for a transform size M, that is large relative to the length of the correlation sequences of the loudspeaker signals, this update equation can be approximated by

$$\underline{w}^N[(k+1)B] = \underline{w}^N[kB] + 2\mu(\mathrm{J}^N 0^{N,M-N})(\breve{R}_x^M[kB])^{-1}(\breve{\mathcal{X}}^M[kB])^T \begin{pmatrix} \underline{0}^{M-B} \\ \underline{r}^B[kB] \end{pmatrix}, \quad (4.7)$$

with $\breve{\mathcal{X}}^M[kB]$ a circulant extension of the data matrix $\mathcal{X}^{N,B}[kB]^T \cdot \mathrm{J}^N$

$$\breve{\mathcal{X}}^M[kB] = \begin{pmatrix} x[kB-M+1] & x[kB] & \dots & x[kB-M+2] \\ \vdots & & & \vdots \\ \vdots & & & x[kB] \\ x[kB] & \dots & \dots & x[kB-M+1] \end{pmatrix}$$

and $\breve{R}_x^M[kB]$ an estimate of the circulant autocorrelation matrix approximation

$$\breve{R}_x^M = \frac{1}{M} E\{(\breve{\mathcal{X}}^M[kB])^T \cdot \breve{\mathcal{X}}^M[kB]\}.$$

Note that the circulant autocorrelation matrix approximation itself is circulant as it is obtained from averaging over the product of the circulant matrices $(\breve{\mathcal{X}}^M)^T$ and $\breve{\mathcal{X}}^M$. The recursive estimation of the approximated autocorrelation matrix can be computed using

$$\breve{R}_x^M[(k+1)B] = \alpha \breve{R}_x^M[kB] + (1-\alpha)((\breve{\mathcal{X}}^M[kB])^T \cdot \breve{\mathcal{X}}^M[kB]), \quad (4.8)$$

with the forgetting factor $0 \leq \alpha < 1$. Usually α chosen close to 1, e.g. $\alpha = 0.99$. For signals that are not stationary a smaller α is used. Note that when choosing α to small, the autocorrelation estimate will become ill conditioned. Pre-multiplying and post-multiplying a right circulant matrix $\mathcal{C}^M$ by the Fourier matrix and its inverse yields a diagonal matrix Davis, 1979

$$\mathcal{F}^M \mathcal{C}^M (\mathcal{F}^M)^{-1} = \mathrm{diag}\{\underline{C}^M\},$$

where the vector $\underline{C}^M$ on the diagonal equals the Fourier transform of the first column of $\mathcal{C}^M$. The advantage of the autocorrelation approximation in (4.8) is that it can be estimated and inverted efficiently in the frequency domain. Pre-multiplying and post-multiplying (4.8) by the Fourier matrix and its inverse yields

$$\mathcal{F}^M \breve{R}_x^M[kB](\mathcal{F}^M)^{-1} = \text{diag}\{\underline{P}^M[kB]\}$$

so that (4.8) becomes

$$\underline{P}^M[(k+1)B] = \alpha \underline{P}^M[kB] + (1-\alpha)\left((\underline{X}^M[kB])^* \otimes \underline{X}^M[kB]\right). \quad (4.9)$$

The inversion of the autocorrelation estimate can now be done efficiently by inverting all diagonal elements element-wise. The inversion of a regular Toeplitz matrix of size $M \times M$ as in (4.2) is prohibited in real-time applications due to the high computational load. Using (4.9) the update equation (4.7) can be transformed efficiently to the frequency domain using the matrix identity $(\breve{\mathcal{F}}^M)^{-1}\breve{\mathcal{F}}^M = \mathrm{I}^M$

$$\begin{aligned}\underline{w}^N &:= \underline{w}^N + 2\mu(\mathrm{J}^N\, 0^{N,M-N})(\breve{\mathcal{F}}^M)^{-1}\breve{\mathcal{F}}^M(\breve{R}_x^M[kB])^{-1}(\breve{\mathcal{F}}^M)^{-1} \\ &\quad \cdot \breve{\mathcal{F}}^M(\breve{\mathcal{X}}^M[kB])^T(\mathcal{F}^M)^{-1}\mathcal{F}^M \begin{pmatrix} \underline{0}^{M-B} \\ \underline{r}^B[kB] \end{pmatrix} \\ &:= \underline{w}^N + 2\mu(\mathrm{J}^N 0^{N,M-N})\,\underline{C}^M[kB], \end{aligned} \quad (4.10)$$

with $\underline{C}^M[kB]$ an estimate of the gradient vector

$$\underline{C}^M[kB] = 2\mu(\underline{P}^M[kB])^{-1} \otimes \underline{R}^M[kB] \otimes \underline{X}^*[kB],$$

with the zero padded and transformed residual signal vector

$$\underline{R}^M[kB] = \mathcal{F}^M \begin{pmatrix} 0^{M-B,B} \\ \mathrm{I}^B \end{pmatrix} \underline{r}^B[kB],$$

and with $(\hat{\underline{P}}^M[kB])^{-1}$ the element-wise inverse of the estimate of the signal power which is defined as

$$\underline{P}^M[kB] = \tfrac{1}{M} E\{\underline{X}^M[kB] \otimes (\underline{X}^M[kB])^*\},$$

which can be computed according to (4.9). This algorithm is the block frequency domain adaptive filter (BFDAF) and is depicted in Figure 4.3. The update part is depicted in Figure 4.4, with D a delay over one block of data.

The multichannel BRLS is introduced in the next chapter. It requires the inversion of a $L_1M \times L_1M$ covariance matrix estimate. This induces a computational load that is much higher than that in the single-channel case. However, by making the same approximation that is done in the present section, the BFDAF is generalized to multichannel BFDAF. The algorithm that is obtained is suitable for real-time operation.

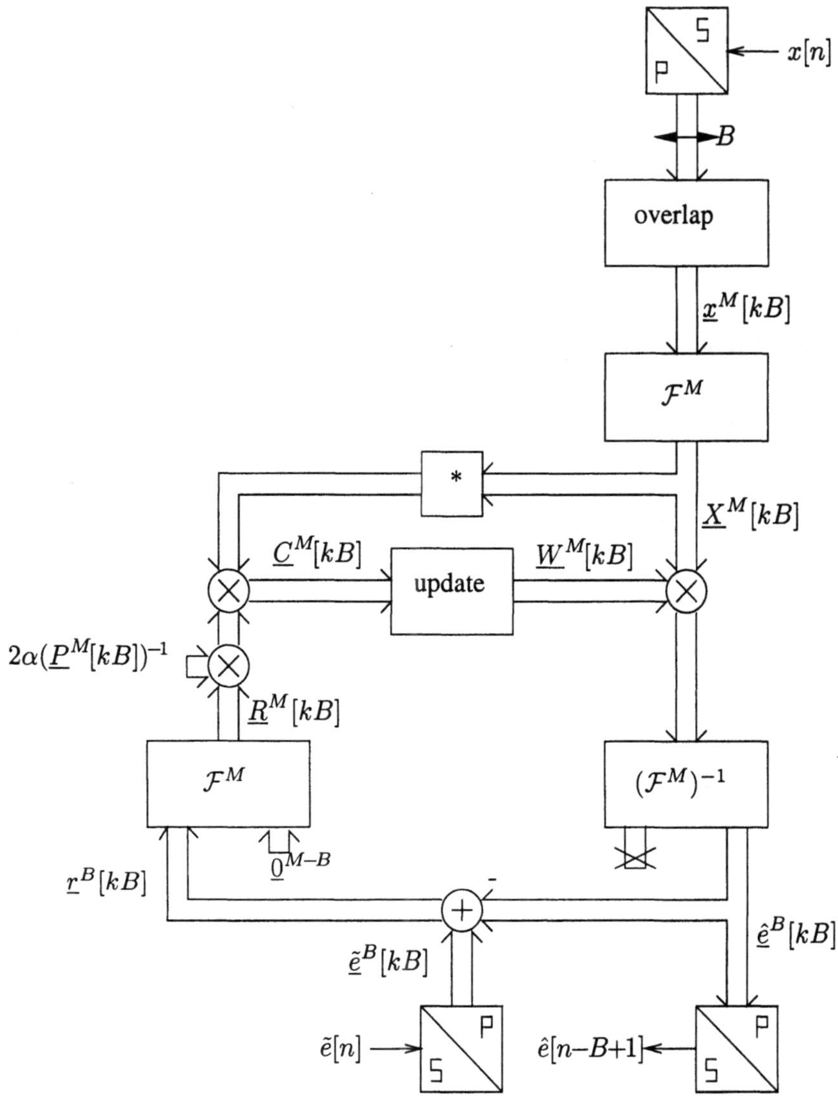

Figure 4.3. Block frequency domain adaptive filter

4.4. Efficient Computation of the FFTs

The computational complexity of a BFDAF is mostly determined by the five (I)FFTs involved, see Sommen, 1992; Mansour and jr., 1982; Egelmeers, 1995, as the filter and update operations have reduced to element-wise multiplications. Although efficient implementations of FFTs have been studied intensively in the past, there is still some improvement that can be achieved by taking into account that some of the (I)FFTs are windowed which is less well known, e.g.

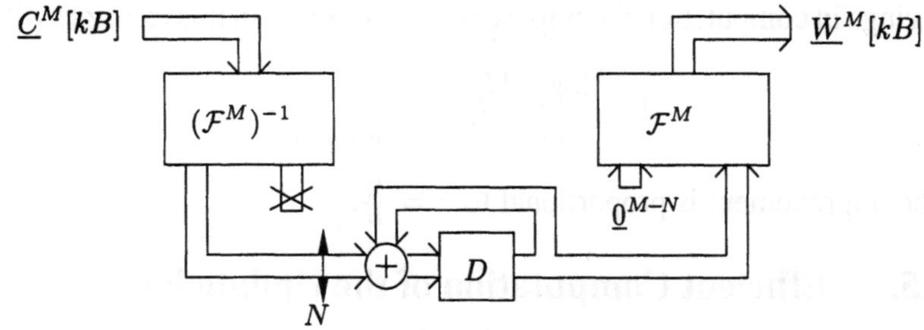

Figure 4.4. BFDAF update part

see Schobben et al., 1997, Egelmeers, 1995 or Appendix D. A windowed FFT transform data to the frequency domain that is appended with zeros and an IFFT transforms data to the time domain of which only a part of the result is required.

The computational complexity of the windowed (I)FFT expressed in the number of multiplication plus the number of additions is about $\frac{5}{2}M\log_2(Q)$ with Q the window size, i.e. the number of samples that are not equal to zero in an FFT or the number of samples that are to be computed in an IFFT.

From Figure 4.3 it can be seen that a windowed FFT can be used to transform the residual signal $\underline{r}^B[kB]$ to the frequency domain. Also, a windowed IFFT can be used to compute the adaptive filter output $\underline{\hat{e}}^B[kB]$. Figure 4.4 shows that the (I)FFTs in the update part are windowed as well. The FFT that transforms the overlapping data $\underline{x}^M[kB]$ to the frequency domain can also be computed with a windowed FFT, see Schobben et al., 1997, Egelmeers, 1995 or Appendix D.

The (I)FFTs in the update part have a window length N while the other three (I)FFTs of the BFDAF have a window length B. When for example $B = N = \frac{M}{2}$, the savings in computational complexity of all (I)FFTs are in this case

$$1 - \frac{\log_2(M/2)}{\log_2(M)} = \frac{1}{\log_2(M)}.$$

For an FFT size of 4096 for example this ratio is about 8% which is significant, and for smaller FFT size the savings in computational complexity are even more pronounced.

In another example, suppose that $B = \frac{M}{r}$ which can be interesting whenever a small algorithm delay is required. The savings in computational complexity are large for the (I)FFTs with window size B. The window size of the (I)FFTs in the update part is N however so that the windowing can not be exploited. In the next section these windows are also reduced to lengths B so that the total savings in computational complexity of all (I)FFTs in this example will become

$$1 - \frac{\log_2(M/r)}{\log_2(M)} = \frac{r}{\log_2(M)}\,.$$

The improvement is proportional to $r = \frac{M}{B}$.

4.5. Efficient Computation of the Update Part

For large adaptive filters in which only a small delay can be tolerated, such as the adaptive echo canceler, B will be small compared to the transform size M. In this case, the filter length N will approximately equal the transform size M. Therefore, the FFT and IFFT in the update part take a dominant amount of flops because the window almost equals the transform size. Furthermore, the efficient implementation of windowed FFTs in Section D.1 requires that both the transform size and the window size are powers of 2 which is not the case. Therefore, 2 regular FFTs are required for the update part. A method is presented next to reduce the computational complexity of the (I)FFT in the update part. The BFDAF update part is shown in Figure 4.4. The update equation (4.10) can be expressed in the frequency domain as

$$\underline{W}^M[(k+1)B] = \mathcal{F}^M \begin{pmatrix} \mathrm{I}^N \\ 0^{N,M-N} \end{pmatrix} \left[\underline{w}^N[kB] + \begin{pmatrix} \mathrm{I}^N & 0^{M-N,N} \end{pmatrix} (\mathcal{F}^M)^{-1} \underline{C}^M[kB]\right]$$

which can be rewritten as

$$\underline{W}^M[(k+1)B] = \underline{W}^M[kB] + \underline{W}^M_\Delta[kB]$$

with

$$\underline{W}^M_\Delta[kB] = \mathcal{F}^M \begin{pmatrix} \mathrm{I}^N & 0^{M-N,N} \\ 0^{M-N,N} & 0^{M-N} \end{pmatrix} (\mathcal{F}^M)^{-1} \underline{C}^M[kB]\,. \tag{4.11}$$

This equation incorporates a time domain window of length N. The computational complexity of the two (I)FFTs in the update part can be reduced by using complementary windowed (I)FFTs. This is achieved by rewriting (4.11)

$$\underline{W}^M_\Delta[kB] = \underline{C}^M[kB] - \mathcal{F}^M \begin{pmatrix} 0^N & 0^{N,M-N} \\ 0^{M-N,N} & \mathrm{I}^{M-N} \end{pmatrix} (\mathcal{F}^M)^{-1} \underline{C}^M[kB]\,. \tag{4.12}$$

The transform size M will now be chosen to be equal to $N + B$, so that the time domain window in (4.12) is of length B similar to the other FFTs in the BFDAF. Consequently $M/2 - 1$ complex additions and 2 real additions are required in (4.12) instead of N real additions. In comparison to (4.11) a reduction is achieved of $\Psi_{\mathrm{FFT}}\{M, M\} - \Psi_{\mathrm{FFT}}\{M, B\} + N - M$ flops. The corresponding modified update is depicted in Figure 4.5.

When using the complementary window all windows are "on the same side" (i.e. the relevant data is always below in the vector) in accordance with Appendix D so that only 1 windowed (I)FFT implementation has to be made. Also, when choosing $M = N + B$, all windows are of the same size.

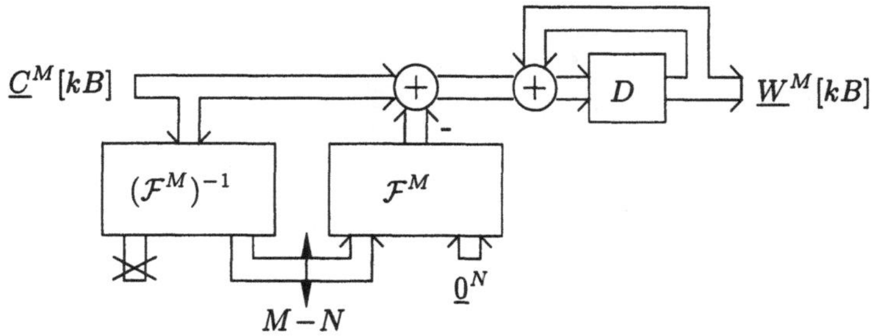

Figure 4.5. Equivalent of BFDAF update part

4.6. Conclusions

A frequency domain adaptive filter contains five (I)FFTs of which four are windowed. A method is presented to implement the windowed (I)FFTs efficiently. The update part is implemented using a complementary time domain window. The overall complexity of the BFDAF is now determined by the five transforms which all have windows of length B. The overall complexity of the proposed implementation of the BFDAF is therefore in the order of $M \log_2(B)/B$ flops per sample, with M the Fourier transform size and B the number of new samples that is used each block. When non-windowed FFTs where used, this complexity would be in the order of $M \log_2(M)/M$ flops. This results in significant computational savings when implementing a large adaptive filter with short a algorithm delay.

Chapter 5

EFFICIENT MULTICHANNEL RLS

This chapter presents a computationally efficient adaptive multichannel algorithm which is derived from the well-known multichannel recursive least squares algorithm. The new algorithm is a generalization of the block frequency domain adaptive filter which is a computationally efficient approximation of the classical recursive least squares algorithm. Application areas in the audio context include stereo echo cancellation for teleconferencing, hands-free telephony and voice-controlled machinery. In such applications, both the desired signal and acoustic echoes that are coming from multiple loudspeakers are picked up by a microphone. Applying multiple single-channel adaptive filters to eliminate such echoes leads to slow convergence when the loudspeaker signals are mutually correlated. This chapter presents a multichannel adaptive filter which uses a joint update rule for all filters. Experiments are carried out to confirm the operation of the multichannel adaptive filter and compare it with the performance of separately updated adaptive filters.

5.1. Introduction

The adaptive filter which is discussed in the previous chapter can be used in applications such as hands-free telephony. The objective is to cancel the contribution of the loudspeaker signal to the microphone signal. The adaptive filter is also known as the acoustic echo canceler (AEC) in this context. When there is a second loudspeaker present that reproduces another sound signal which is independent of the first one, a second AEC can be used to eliminate its contribution to the microphone signal. Problems arise however when the loudspeaker signals are not independent of each other. This occurs for example in applications that involve stereophonic sound reproduction. In that case, applying multiple AECs independently of each other gives rise to slow convergence properties of these adaptive filters as the update of one filter depends on that of the others.

In this chapter a joint update algorithm is derived that controls multiple filters. Its performance is superior to using separate updates when filtering signals that exhibit strong mutual correlations.

In single channel LMS, the convergence speed is governed by the eigenvalue spread of the autocorrelation matrix. In RLS this problem is remedied by pre-multiplying the update term with the inverse of the autocorrelation matrix. The computationally efficient block frequency domain adaptive filter (BFDAF) from the previous chapter remedies this problem by applying a normalization in the frequency domain. This is generalized to the multichannel case in this chapter.

Furthermore, in multichannel adaptive filters also the problem arises that the filters are not unique. There is a large set of filters that can set the residual error equal to zero when only one speaker is active in the far-end room of a teleconferencing setup. It is desirable however to correctly identify the unknown echo paths in the near-end room. The adaptation algorithm only has to track changes in these echo paths in that case which are only influenced by speakers that walk through the near-end room. When the adaptive algorithm does not correctly identify the echo paths, it must also track variations in the far-end room. Even when the people in the far end room are not moving, re-convergence is needed whenever one speaker stops talking and another speaker starts talking at another location as described in Sondhi et al., 1995. The uniqueness problem can be remedied by preprocessing the loudspeaker signals so that the cross-correlations among the channels vary, see Hirano et al., 1999; Joncour and Sugiyama, 1998; Ali, 1998; Benesty et al., 1998; Shimauchi et al., 1999. These variations must be such that they are inaudible or at least non-disturbing. Another approach is to add independent contributions to the loudspeaker signals such as noise signals, see Gillore and Turbin, 1998; Shimauchi et al., 1998. Again, such contributions must not be disturbing to the listener. A more detailed introduction on multichannel AEC can be found in Sondhi et al., 1995; Benesty et al., 1995.

In this chapter an approach is followed which accommodates for multichannel echo cancellation at a computational cost that is only slightly higher than when single channel AECs are applied. It is shown that it can be implemented in real-time on a typical desktop computer. This is not possible for classical multichannel RLS, where the size of the cross-correlation matrix grows linearly with the number of channels times the number of filter coefficients per channel. Estimating and inverting such a matrix is not possible in real-time. An LMS like algorithm is obtained when this cross-correlation matrix is not used to normalize the adaptive algorithm resulting in a poor convergence. Other methods have been reported which aim on diagonalizing the multichannel cross-correlation matrix based on transform domain and sub-band adaptive filtering as in Alberge et al., 1998.

An application of the multichannel AEC is depicted in Figure 5.1. Here,

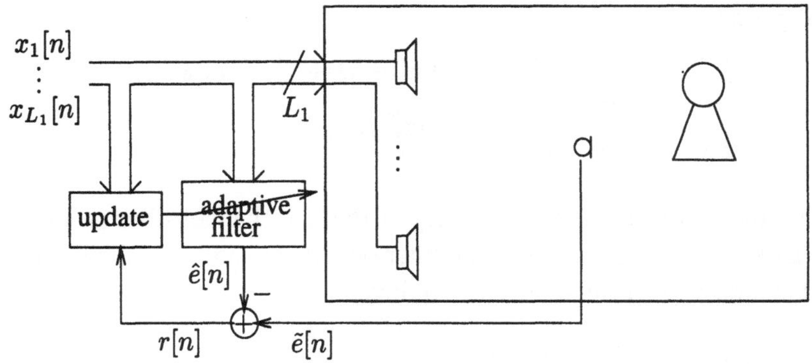

Figure 5.1. Multichannel acoustic echo cancellation

sounds that are produced by the L_1 loudspeakers are picked up by the microphone together with the speech of the local speaker. The multichannel adaptive filter aims at estimating the contribution of the L_1 loudspeaker signals to the microphone signal by minimizing the difference between this estimate $\hat{e}[n]$ and the microphone signal $\tilde{e}[n]$. Multiple microphones are used in applications such as multichannel teleconferencing. The same approach can be followed for all microphone signals independently in that case. Canceling the contribution of multiple loudspeakers to multiple microphones is therefore trivial and the algorithm will be derived therefore for multiple loudspeakers and a single microphone. Note that similar to the classic single-channel adaptive filter, a double talk detector is needed which detects if the local speaker is active. If so, the adaptive filters are kept constant or they are updated more slowly. It is well-known that the performance of the echo canceler deteriorates otherwise.

5.2. Filter Part

Computationally efficient filter implementations are obtained using block processing in the frequency domain as in the previous chapter. Further improvements in computational efficiency can be obtained for applications in which only short processing delays are allowed. In such applications, the adaptive filter can be partitioned into smaller length sub-filters, see Sommen, 1989.

The filtering involved in a multiple-input single-output FIR filter is achieved by filtering all inputs efficiently in the frequency domain as in (4.4) and summing up the results

$$\underline{\hat{e}}^B[kB] = (0^{B,M-B}\; I^B)\cdot(\mathcal{F}^M)^{-1}\cdot\left(\sum_{j=1}^{L_1}\left(\underline{X}_j^M[kB]\otimes\underline{W}_j^M[kB]\right)\right)$$

with $B \geq 1$ and

$$\begin{aligned} \underline{X}_j^M[kB] &= \mathcal{F}^M \cdot (x_j[kB-M+1]\ldots x_j[kB])^T \\ \underline{W}_j^M[kB] &= \mathcal{F}^M \cdot \begin{pmatrix} \mathrm{J}^N \underline{w}_j^N[kB] \\ \underline{0}^{M-N} \end{pmatrix}. \end{aligned}$$

Note that the computational complexity of this L_1 channel filter is less than that of L_1 single channel filters as the filter outputs are added in the frequency domain and only one inverse FFT is required.

5.3. Multichannel RLS Update

The update equation of single channel RLS is generalized to the multichannel case in this section. The multichannel RLS update is obtained by stacking the filter vectors and data matrices in (4.2)

$$\underline{w}^{L_1 N} := \underline{w}^{L_1 N} + 2\mu(\hat{R}_x^{L_1 N})^{-1}\mathcal{X}^{L_1 N,B}[kB]\underline{r}^B[kB], \tag{5.1}$$

with $\hat{R}_x^{L_1 N}$ an estimate of the multichannel autocorrelation matrix $R_x^{L_1 N}$, with

$$\begin{aligned} \mathcal{X}^{L_1 N,B}[kB] &= \begin{pmatrix} \mathcal{X}_1^{N,B}[kB] \\ \vdots \\ \mathcal{X}_{L_1}^{N,B}[kB] \end{pmatrix}, \\ R_x^{L_1 N} &= E\{\mathcal{X}^{L_1 N,B}[kB](\mathcal{X}^{L_1 N,B}[kB])^T\}, \\ \underline{w}^{L_1 N} &= \begin{pmatrix} \underline{w}_1^N \\ \vdots \\ \underline{w}_{L_1}^N \end{pmatrix}. \end{aligned}$$

Note that in the multichannel RLS update all data matrices are multiplied with the same residual signal $\underline{r}^B[kB]$. Both the autocorrelations and the correlations between the various channels are removed by pre-multiplying with the inverse of an estimate of the covariance matrix that is based on data from all channels. The same approximation that was used in the previous section for the autocorrelation matrix is used here for the covariance matrix in the multichannel update so that the update can be written as

$$\underline{w}^{L_1 N} := \underline{w}^{L_1 N} + 2\mu \mathrm{J}_N^{L_1 N, L_1 M}(\breve{R}_x^{L_1 M}[kB])^{-1} \cdot (\breve{\mathcal{X}}^{M, L_1 M}[kB])^T \begin{pmatrix} \underline{0}^{M-B} \\ \underline{r}^B[kB] \end{pmatrix} \tag{5.2}$$

with $\mathrm{J}_N^{L_1 N, L_1 M}$ the multichannel windowed mirror matrix

$$\mathrm{J}_N^{L_1 N, L_1 M} = \begin{pmatrix} \mathrm{J}^N & & & 0^{N, L_1 M - N} \\ 0^{N,M} & \mathrm{J}^N & & 0^{N,(L_1-1)M-N} \\ & & \ddots & \\ 0^{N,(L_1-1)M} & & \mathrm{J}^N & 0^{N,M-N} \end{pmatrix}$$

and $\breve{\mathcal{X}}^{M,L_1M}[kB]$ the block-circulant multichannel data matrix

$$(\breve{\mathcal{X}}^{M,L_1M}[kB])^T = \begin{pmatrix} (\breve{\mathcal{X}}_1^M[kB])^T \\ \vdots \\ (\breve{\mathcal{X}}_{L_1}^M[kB])^T \end{pmatrix}$$

and $\hat{\breve{R}}_x^{L_1M}[kB]$ an estimate of the block-circulant multichannel covariance matrix approximation $\breve{R}_x^{L_1M}$

$$\breve{R}_x^{L_1M} = \frac{1}{M} E\{(\bar{\mathcal{X}}^{M,L_1M}[kB])^T \bar{\mathcal{X}}^{M,L_1M}[kB]\},$$

with the block circulant data matrix

$$(\bar{\mathcal{X}}^{M,L_1M}[kB])^T = \begin{pmatrix} \breve{\mathcal{X}}_1^M[kB] \\ \vdots \\ \breve{\mathcal{X}}_{L_1}^M[kB] \end{pmatrix}.$$

The multichannel update equation (5.2) can be transformed to the frequency domain by pre and post-multiplying by the multichannel Fourier matrix $\breve{\mathcal{F}}^{L_1M}$ and its inverse

$$\begin{aligned} \underline{w}^{L_1N} := \underline{w}^{L_1N} &+ 2\mu \mathrm{J}_N^{L_1N,L_1M} (\breve{\mathcal{F}}^{L_1M})^{-1} \cdot \breve{\mathcal{F}}^{L_1M} (\hat{\breve{R}}_x^{L_1M}[kB])^{-1} (\breve{\mathcal{F}}^{L_1M})^{-1} \\ &\cdot \breve{\mathcal{F}}^{L_1M} (\breve{\mathcal{X}}^{M,L_1M}[kB])^T (\mathcal{F}^M)^{-1} \cdot \mathcal{F}^M \begin{pmatrix} \underline{0}^{M-B} \\ \underline{r}^B[kB] \end{pmatrix}, \end{aligned} \tag{5.3}$$

with the multichannel Fourier transform matrix

$$\breve{\mathcal{F}}^{L_1M} = \begin{pmatrix} \mathcal{F}^M & 0^M & \dots & 0^M \\ 0^M & \mathcal{F}^M & & \vdots \\ \vdots & & \ddots & 0^M \\ 0^M & \dots & 0^M & \mathcal{F}^M \end{pmatrix}.$$

Note that $(\breve{\mathcal{F}}^{L_1M})^{-1}$ has inverse Fourier matrices on its main diagonal and zeros elsewhere. For an efficient implementation of (5.3), the block diagonal data matrix is rewritten as

$$\breve{\mathcal{F}}^{L_1M} \cdot (\breve{\mathcal{X}}^{M,L_1M}[kB])^T \cdot (\mathcal{F}^M)^{-1} = \begin{pmatrix} \mathrm{diag}\{(\underline{X}_1^M[kB])^*\} \\ \vdots \\ \mathrm{diag}\{(\underline{X}_{L_1}^M[kB])^*\} \end{pmatrix}. \tag{5.4}$$

A similar diagonalization is used for $\bar{\mathcal{X}}^{M,L_1M}[kB]$

$$(\breve{\mathcal{F}}^{L_1M})^{-1} \cdot (\bar{\mathcal{X}}^{M,L_1M}[kB])^T \cdot \mathcal{F}^M = \begin{pmatrix} \mathrm{diag}\{(\underline{X}_1^M[kB])^*\} \\ \vdots \\ \mathrm{diag}\{(\underline{X}_{L_1}^M[kB])^*\} \end{pmatrix},$$

so that the covariance matrix estimate can be calculated efficiently in the frequency domain by expanding the approach in (4.9) to the multichannel case

$$\begin{aligned}
&\breve{\mathcal{F}}^{L_1M}(\breve{R}_x^{L_1M}[kB])^{-1}(\breve{\mathcal{F}}^{L_1M})^{-1}\\
&=\left((\breve{\mathcal{F}}^{L_1M})^{-1}\breve{R}_x^{L_1M}[kB]\breve{\mathcal{F}}^{L_1M}\right)^{-1}\\
&=\left(\frac{1}{M}\hat{E}\left\{(\breve{\mathcal{F}}^{L_1M})^{-1}(\bar{\mathcal{X}}^{M,L_1M}[kB])^T\mathcal{F}^M(\mathcal{F}^M)^{-1}\bar{\mathcal{X}}^{M,L_1M}[kB]\breve{\mathcal{F}}^{L_1M}\right\}\right)^{-1}\\
&=\left(\frac{1}{M}\hat{E}\left\{\begin{pmatrix}\mathrm{diag}\{(\underline{X}_1^M)^*\}\\ \vdots\\ \mathrm{diag}\{(\underline{X}_{L_1}^M)^*\}\end{pmatrix}\left(\mathrm{diag}\{\underline{X}_1^M\}\dots\mathrm{diag}\{\underline{X}_{L_1}^M\}\right)\right\}\right)^{-1}\\
&=\begin{pmatrix}\mathrm{diag}\{\underline{P}_{11}^M[kB]\} & \dots & \mathrm{diag}\{\underline{P}_{1L_1}^M[kB]\}\\ \vdots & \ddots & \vdots\\ \mathrm{diag}\{\underline{P}_{L_11}^M[kB]\} & \dots & \mathrm{diag}\{\underline{P}_{L_1L_1}^M[kB]\}\end{pmatrix}^{-1}, \qquad (5.5)
\end{aligned}$$

with $\hat{E}\{.\}$ the averaging operator that is used to recursively estimate the cross-power vectors $\underline{P}_{ac}^M$ for all a, c

$$\underline{P}_{ac}^M[(k+1)B] = \alpha\underline{P}_{ac}^M[kB] + (1-\alpha)\left((\underline{X}_a^M[kB])^* \otimes \underline{X}_c^M[kB]\right).$$

Note that the elements of $\underline{P}_{ac}^M$ are complex conjugate symmetrical as $\underline{X}_a^M[kB]$ and $\underline{X}_c^M[kB]$ are transforms of real data. Therefore, only half of the elements need to be calculated. When $a = c$, the elements of the cross-power vector $\underline{P}_{ac}^M$ are real so that the complex parts do not need to be calculated. Moreover, $\underline{P}_{ac}^M = (\underline{P}_{ca}^M)^*$ so only one of each two cross-power vectors need to be calculated. The results in (5.4) and (5.5) are substituted in the update equation (5.3)

$$\begin{aligned}
&\underline{w}^{L_1N} := \underline{w}^{L_1N} + 2\mu \mathrm{J}_N^{L_1N,L_1M}(\breve{\mathcal{F}}^{L_1M})^{-1}\\
&\cdot\begin{pmatrix}\mathrm{diag}\{\underline{P}_{11}^M[kB]\} & \dots & \mathrm{diag}\{\underline{P}_{1L_1}^M[kB]\}\\ \vdots & \ddots & \vdots\\ \mathrm{diag}\{\underline{P}_{L_11}^M[kB]\} & \dots & \mathrm{diag}\{\underline{P}_{L_1L_1}^M[kB]\}\end{pmatrix}^{-1}\cdot\begin{pmatrix}\mathrm{diag}\{(\underline{X}_1^M)^*\}\\ \vdots\\ \mathrm{diag}\{(\underline{X}_{L_1}^M)^*\}\end{pmatrix}\cdot\underline{R}^M[kB] \qquad (5.6)
\end{aligned}$$

with

$$\underline{R}^M[kB] = \mathcal{F}^M\begin{pmatrix}\underline{0}^{M-B}\\ \underline{r}^B[kB]\end{pmatrix}.$$

The block-diagonal frequency domain covariance matrix can now be inverted efficiently due to its sparse structure. The efficient calculation of this inverse is shown in the next section. The update equation in (5.6) together with the multichannel filter that is controlled by it is a generalization of the block frequency domain adaptive filter (BFDAF). This generalization will be termed the multichannel block frequency domain adaptive filter (MC-BFDAF).

5.4. Inverting the Covariation Matrix

It was shown in the previous section that the covariation matrix can be made block diagonal. In this section it is shown that the inverse of a block diagonal matrix is again block diagonal and a method is presented to calculate such an inverse matrix. For notational convenience, let the sub-matrices of the matrix to be inverted in (5.5) and those of the inverse matrix be denoted as $\underline{D}_{kl}$ and $\underline{E}_{kl}$ respectively. The sub-matrices of the inverse matrix $\underline{E}_{kl}$ can be found from

$$\begin{pmatrix} \mathrm{diag}\{\underline{D}_{11}^M\} & \dots & \mathrm{diag}\{\underline{D}_{1L_1}^M\} \\ \vdots & \ddots & \vdots \\ \mathrm{diag}\{\underline{D}_{L_1 1}^M\} & \dots & \mathrm{diag}\{\underline{D}_{L_1 L_1}^M\} \end{pmatrix} \cdot \begin{pmatrix} \mathrm{diag}\{\underline{E}_{1}^M\} & \dots & \mathrm{diag}\{\underline{E}_{1L_1}^M\} \\ \vdots & \ddots & \vdots \\ \mathrm{diag}\{\underline{E}_{L_1 1}^M\} & \dots & \mathrm{diag}\{\underline{E}_{L_1 L_1}^M\} \end{pmatrix} = \mathrm{I}^{L_1 M}.$$

This set of equations can be rearranged so that for $m = 1 \dots L_1$:

$$\begin{pmatrix} (\underline{D}_{11}^M)_m & \dots & (\underline{D}_{1L_1}^M)_m \\ \vdots & \ddots & \vdots \\ (\underline{D}_{L_1 1}^M)_m & \dots & (\underline{D}_{L_1 L_1}^M)_m \end{pmatrix} \cdot \begin{pmatrix} (\underline{E}_{11}^M)_m & \dots & (\underline{E}_{1L_1}^M)_m \\ \vdots & \ddots & \vdots \\ (\underline{E}_{L_1 1}^M)_m & \dots & (\underline{E}_{L_1 L_1}^M)_m \end{pmatrix} = \mathrm{I}^{L_1}.$$

So, the inversion of the $L_1 M \times L_1 M$ block diagonal matrix is identical to the inversion of M matrices of size $L_1 \times L_1$. The cross-power estimates $(\underline{P}_{kl}^M)_m$ are real for $m = 0$ and $m = \frac{M}{2}$ and complex symmetrical for $m = 1 \dots \frac{M}{2} - 1$, i.e. $(\underline{P}_{kl}^M)_{M-m} = ((\underline{P}_{kl}^M)_m)^*$. Therefore, the inversion of the $L_1 M \times L_1 M$ block diagonal covariance matrix corresponds to the inversion of 2 real $L_1 \times L_1$ matrices and $\frac{M}{2} - 1$ complex $L_1 \times L_1$ matrices.

5.5. Algorithm Overview

In this section an overview is given of the multichannel block frequency domain adaptive filter. The adaptive procedure consists of the following steps;

1 Blocks of microphone signals are transformed to the frequency domain for $a = 1 \dots L_1$

$$\underline{X}_a^M[kB] = \mathcal{F}^M \begin{pmatrix} x_a[kB - M + 1] \\ \vdots \\ x_a[kB] \end{pmatrix}.$$

The blocks are of length M and are overlapping; only B new samples are used per block.

2 The cross-power estimates are updated efficiently in the frequency domain $\forall a, c : \underline{P}_{ac}^M := \alpha \underline{P}_{ac}^M + (1 - \alpha)\left((\underline{X}_a^M[kB])^* \otimes \underline{X}_c^M[kB]\right)$

The forgetting factor α may vary from 0 to 1 depending on the application. Usually α is chosen near to 1, e.g. $\alpha = 0.99$ which is optimal (nearly) stationary signals.

3 The cross-power matrix is inverted efficiently as shown in Section 5.4. The inverse of the cross-power matrix has the same sparse structure as the cross-power matrix itself and is denoted as $\begin{pmatrix} \mathrm{diag}\{\underline{Q}_{11}^M\} & \dots & \mathrm{diag}\{\underline{Q}_{1L_1}^M\} \\ \vdots & \ddots & \vdots \\ \mathrm{diag}\{\underline{Q}_{L_1 1}^M\} & \dots & \mathrm{diag}\{\underline{Q}_{L_1 L_1}^M\} \end{pmatrix}$.

4 The filters are updated for $a = 1 \dots L_1$

$$\underline{w}_a^N := \underline{w}_a^N + 2\mu(\mathrm{J}^N \, 0^{N,M-N}) \cdot (\breve{\mathcal{F}}^M)^{-1} \sum_{c=1}^{L_1} \underline{Q}_{ac}^M \otimes (\underline{X}_c^M[kB])^* \otimes \underline{R}^M[kB]$$

5 The filtering is performed efficiently in the frequency domain using the overlap-save method

$$\underline{\hat{e}}^B[kB] = (0^{B,M-B} \; \mathrm{I}^B)(\mathcal{F}^M)^{-1} \sum_{a=1}^{L_1} \underline{X}_a^M[kB] \otimes \underline{W}_a^M[kB],$$

with $\underline{W}_a^M[kB] = \mathcal{F}^M \cdot \begin{pmatrix} \mathrm{J}^N \underline{w}_a^N[kB] \\ \underline{0}^{M-N} \end{pmatrix}$.

6 The residual error is calculated and transformed to the frequency domain

$$\underline{R}[kB] = \mathcal{F}^M \begin{pmatrix} 0^{M-B} \\ \breve{e}[kB] - \hat{e}[kB] \end{pmatrix}.$$

7 All steps are repeated iteratively.

Note that the filtering and the update can be calculated independently. Reducing the update rate lowers the computational complexity at the cost of a slower convergence.

5.6. Computational Complexity

The computational complexity of the frequency domain multichannel block frequency domain adaptive filter is estimated in this section. In the L_1 channel MC-BFDAF algorithm, L_1 FFTs are needed to transform the input data and 1 IFFT is needed to transform the filter output back to the time domain. In the update, L_1 IFFTs and L_1 FFTs are needed to transform the filter coefficients back and forth from the time domain to the frequency domain. Also, 1 FFT is needed to transform the residual signal $\underline{r}^B[kB]$. All these $3L_1 + 2$ (I)FFTs are windowed meaning that they are used to transform zero padded data or inverse transform data from which only a part is needed. This requires approximately $(3L_1 + 2)\frac{5}{2}M\log_2(B)$ floating point operations as mentioned in the previous chapter.

The $\frac{M}{2}$ $L_1 \times L_1$ complex matrix inversions that are used to invert the covariance matrix take in the order of $M{L_1}^3$ operations. In the efficient implementation of these matrix inversions the property can be used that they are complex symmetrical, i.e. complex conjugated transposition results in the same matrix.

In the filter part, L_1 element-wise vector products are used. In the update, L_1 element-wise vector products are needed to multiply the data with the residual vector and another ${L_1}^2$ element-wise products are needed to multiply this result with the covariance matrix inverse. In addition, $\frac{{L_1}^2+L_1}{2}$ element-wise vector products are needed to update the covariance matrix estimate. In total, $\frac{3}{2}{L_1}^2 + \frac{5}{2}L_1$ element-wise products are needed which corresponds to $3M(\frac{3}{2}{L_1}^2 + \frac{5}{2}L_1)$ real floating point operations.

The overall computational complexity that is needed to produce a block of B output samples is approximately given by

$$\Psi_{\text{MCBFDAF}}\{B, M, L_1\} = (3L_1 + 2)\frac{5}{2}M\log_2(B) + M{L_1}^3 + 3M(\frac{3}{2}{L_1}^2 + \frac{5}{2}L_1)$$

floating point operations. Evaluating this complexity for 3 channels ($L_1 = 3$), length 1024 FFTs ($M = 1024$), a sample frequency of 44.1 kHz and an algorithm delay of 6 ms ($B = 256$) yields a complexity of 55 megaflops. Although there will be additional overhead when implemented, real-time operation is feasible as a typical desktop computer can do about 1000 megaflops.

For a small number of channels ($L_1 \leq 4$), the overall complexity is determined by the (I)FFTs. When there are many channels ($L_1 > 6$), the inversion of the $L_1 \times L_1$ matrices determines the overall complexity.

5.7. Experiments

The MC-BFDAF algorithm that is introduced in this chapter has been implemented in Matlab. In the experiments there is no active local speaker so that double talk is not a problem. An important application of the MC-BFDAF is canceling the contribution of highly correlated stereo sounds generated by loudspeakers to the signal that is picked up by a microphone in the same room as is depicted in Figure 5.1. The benefit of the multichannel normalization in MC-BFDAF can be measured by comparing it against multiple separate single channel BFDAFs. A comparison with the time domain MC-BRLS in (5.1) cannot be made for acoustic applications that require long filters due to limited memory and computation time. This comparison would also be less interesting from an application point of view as implementing MC-BRLS in real-time is not feasible.

In this experiment a stereo recording of two people that are speaking simultaneously is synthetically mixed using two measured impulse responses h_1 and h_2 which are plotted in Figure 5.2. The resulting 'microphone' signal is used as the reference signal for the MC-BFDAF. The task of the MC-BFDAF is now to make an estimate of this reference signal by filtering the stereo speech tracks. The two FIR filters that are controlled by the MC-BFDAF both have 1024 tabs. The forgetting factor in the cross-power estimation is $\alpha = 0.9$. The number of new samples that is processed each block is 1024 in this experiment. A Fourier transform size of 2048 is used to transform the data to the frequency domain. The sample frequency is 16 kHz. The audio tracks that are processed are of 10 seconds duration. In Figure 5.3 both the microphone signal and the residual signal after stereo cancellation are plotted as a function of number of samples that is processed. It can be seen from this plot that the MC-BFDAF converges and that the residual signal goes to zero accordingly. In Figure 5.4 the power of the microphone signal and the residual signal are plotted as a function of number of samples that is processed. After 10 seconds, a suppression of about 32 dB is obtained.

In the next experiment the same parameters are used as in the previous experiment but the update of the two filters is now done separately. Both the microphone signal and the resulting residual signal are plotted in Figure 5.5. When comparing the residual signal with that of the previous experiment it can be seen that the convergence properties of MC-BFDAF are superior to that of multiple single-channel BFDAFs. Again, the power of the microphone signal and the residual signal are plotted as a function of the number of samples that is processed in Figure 5.6. Only about 17 dB suppression is obtained after 10 seconds in this case. As opposed to the previous experiment, the far-end signals are still clearly audible in the residual signal in this case.

The human ear is the final judge when determining quality of sound. The sounds from the experiments in this section and the Matlab code of the MC-BFDAF are therefore included on compact disc. A description of the compact disc is given in Appendix E.

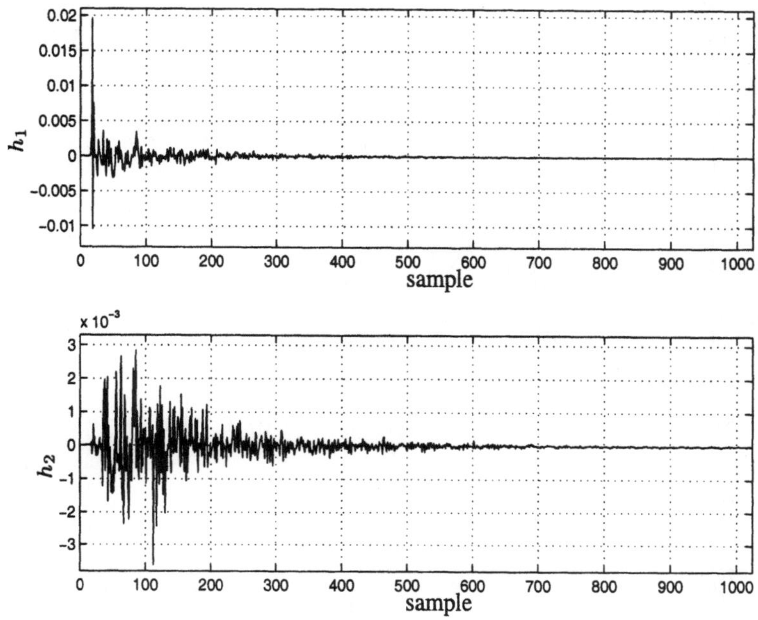

Figure 5.2. Acoustic impulse responses from the sound sources to the microphone

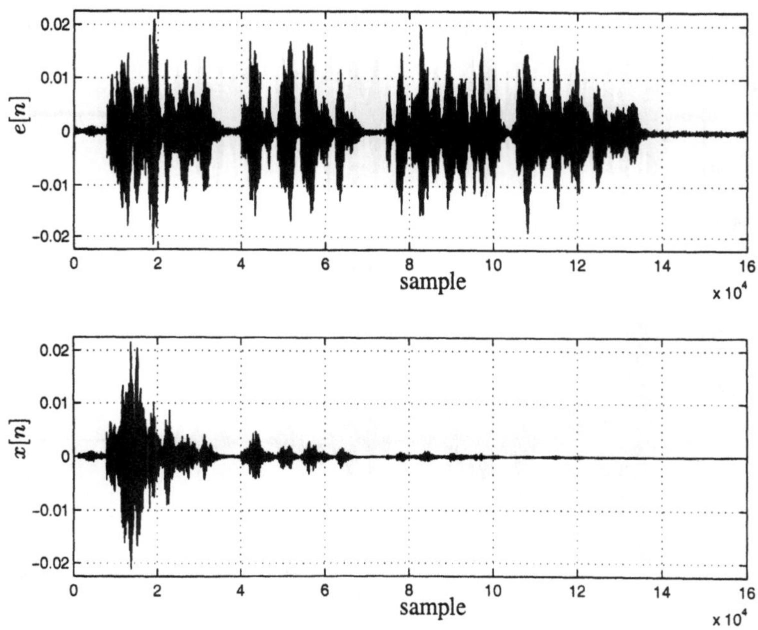

Figure 5.3. Microphone signal (*top*) and residual signal (*bottom*) after stereo echo cancellation

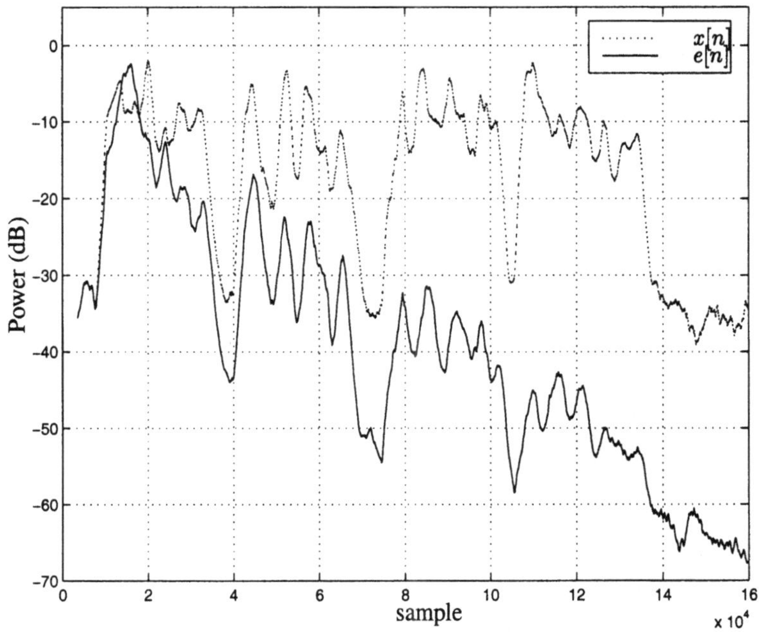

Figure 5.4. Power of the microphone signal and the residual signal after stereo echo cancellation

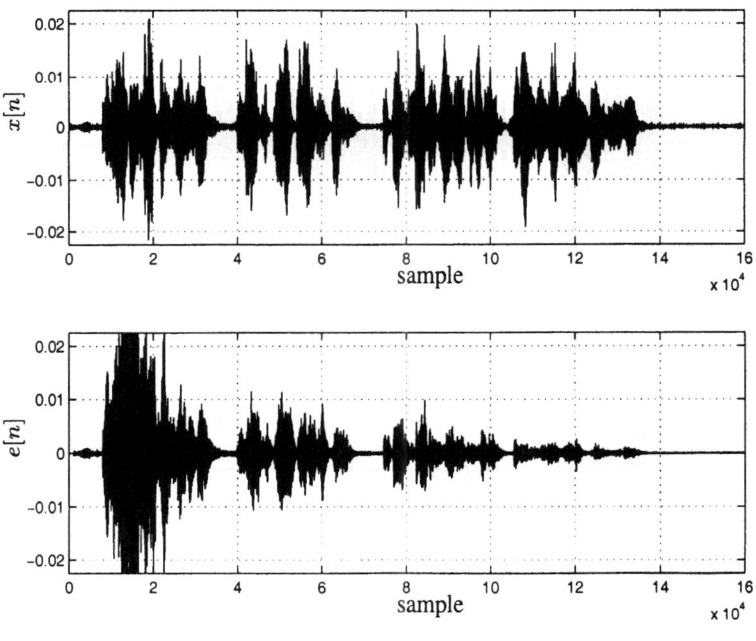

Figure 5.5. Microphone signal (*top*) and residual signal (*bottom*) using two independent echo cancelers

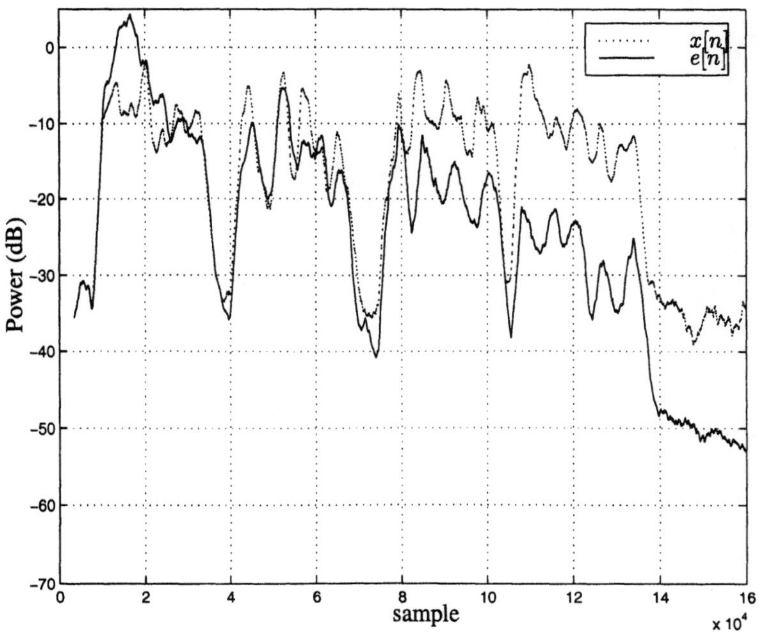

Figure 5.6. Power of the microphone signal and the residual signal using two independent echo cancelers

III

BLIND SIGNAL SEPARATION

Chapter 6

BLIND SIGNAL SEPARATION, AN OVERVIEW

Blind Signal Separation (BSS) is the process of recovering independent signals that correspond to the individual source signals using only observed linear mixtures of these. In an acoustic context, these source signals are correlated in time and are assumed to be independent of each other. The mixing system is convolutive in the sense that the microphones pick up delayed and attenuated versions of the source signals due to reflections in the room. The microphone signals typically contain some microphone noise.

As discussed briefly in Section 2.2, BSS is a special case of blind deconvolution. Blind deconvolution comprises both signal separation and signal equalization; i.e. it aims at recovering the exact sound sources. In an acoustic context this means that it tries to perform BSS and dereverberation simultaneously. Although BSS is not concerned with dereverberation, is it useful to put constraints into the BSS system so that its output signals sound natural.

This chapter gives an overview of existing methods for BSS. It is apparent from this overview that many algorithms reside to higher order statistics (HOS) to achieve BSS. Until recently, many researchers assumed that second order statistics are insufficient to achieve signal separation. This chapter concludes with an explanation of why this is not the case. A new BSS algorithm is presented in the next chapter that is based entirely on second order statistics.

6.1. Overview of Existing Methods

A BSS algorithm consists of a learning rule and the corresponding filter structure. These learning rules come from various different approaches that use different assumptions on the mixing system and use different properties of the source signals. At a more detailed level, any BSS algorithm can be characterized by

- The assumptions that are made on the signals or the mixing system. Assumptions are made typically on the source signals and include mutual independence and (non-)stationarity. Assumptions that are made on the mixing system include that the individual transfer functions are minimum-phase and that the individual transfer functions are of finite length.

- The criterion that is optimized. This includes minimizing the mutual entropy among the outputs, minimizing the cross-correlations among the outputs of the unmixing system, maximizing the distance between the probability density function (PDFs) of all output signals and a Gaussian PDF and minimizing cross-HOS among the outputs.

- The parameterization of the signals or the mixing system that is used in this criterion. This includes autoregressive modeling of the sources, modeling of the sources' PDFs and modeling of the direct transmission paths from the sources to the microphones (delay estimation). Note that some algorithms do not require parameterization.

- The method that is used to optimize the criterion. This method often is a gradient descent method. Important is that the method operates on parameters in the unmixing system that are as independent as possible. This is required to achieve fast convergence especially in large unmixing systems. Frequency domain approaches have shown to be promising for this purpose.

- The filter structure of the unmixing system. Both feed-forward and feedback systems have been presented in the literature. Feedback systems have the ability to closer resemble of the exact inverse of the mixing system in some cases. A disadvantage is the inherent stability problem. Also, feed-forward systems have the advantage that they can be implemented more efficiently in the frequency domain using the overlap-save technique.

BSS algorithms have been successful for separating non-convolutive mixtures of non-real-world signals for over a decade. Successful BSS of non-convolutive mixed, delayed mixed and synthetical convolutive mixed real audio signals were reported in Bell and Sejnowski, 1995b; Torkkola, 1996; Weinstein et al., 1993 respectively. Only after 1995 successful experiments are reported with the comprehensive problem of separating signals that are recorded using microphones in a real world environment, as in Lee et al., 1997a; Lambert and Bell, 1997; Smaragdis, 1998; Ikeda and Murata, 1999; Parra and Spence, 1998; Sahlin and Broman, 1997; v. Gerven, 1996; Chan, 1997; Schobben and Sommen, 1998a. Most of these experiments are done in controlled laboratory environments and require improvements to be successful in commercial applications.

An overview of the early work of non-convolutive and convolutive BSS can be found in Pope and Bogner, 1996a and Pope and Bogner, 1996b respectively.

More recent work is published in Haykin, 2000b. An overview of BSS approaches that are promising for audio applications is found in Torkkola, 1999 and in the next sections. In Joho, 2000 blind signal separation is studied by starting from echo cancellation methods which are non-blind in the sense that they rely on a reference signal.

The cascaded mixing and unmixing of two sources is depicted in Figure 6.1. In many publications, the mixing and unmixing process is simplified by

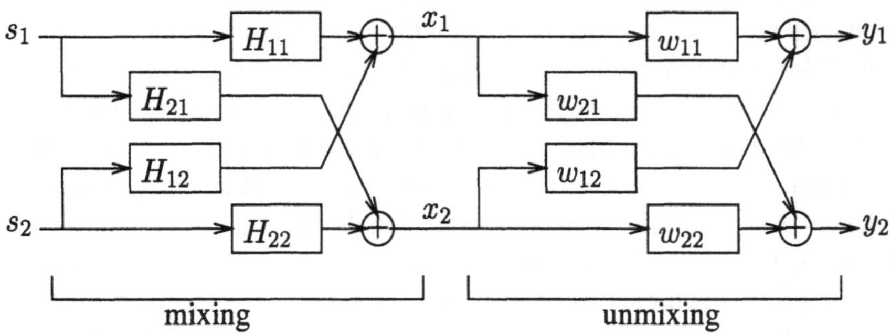

Figure 6.1. Cascaded mixing/unmixing of two sources

assuming that $H_{11} = H_{22} = 1$ and using $w_{11} = w_{22} = 1$. There exist two solutions for the unmixing filters to achieve signal separation in that case

$$\left\{ \begin{array}{l} w_{12} = -H_{12} \\ w_{21} = -H_{21} \end{array} \right. \quad \text{and} \quad \left\{ \begin{array}{l} w_{12} = -\frac{1}{H_{21}} \\ w_{21} = -\frac{1}{H_{12}} \end{array} \right. . \tag{6.1}$$

These solutions both lead to signal separation but recover the source signals at different outputs. Note that the direct filters H_{ii} can always be absorbed by the sources so that this model remains valid. This corresponds to considering the contribution of the ith source to the ith microphone as the source in the model. Assuming one is interested in BSS and not in dereverbation, this is a suitable model.

Most algorithms presented in the literature deal with recovering as many signals as there are sensors and many of them are restricted to recovering 2 sources from 2 sensors. Many algorithms cannot be extended in a straightforward manner to deal with more than 2 sources and sensors or to deal with more sensors than sources. It is apparent from Section 2.4 that using more sensors than sources is promising when the filter structure that is used for unmixing is a multichannel FIR filter.

Output Decorrelation

The objective of output decorrelation is to produce uncorrelated signals. This is typically achieved by minimizing a cost function that is composed of cross-correlations among the outputs, e.g.

$$\min\{E\{y_i[n]y_j[n-\tau]\}\}, \; i \neq j.$$

Molgedey and Schuster minimized cross-correlations for two different time-lags to achieve signal separation in the instantaneous case in Molgedey and Schuster, 1994. Multiple time-lags are used for convolutive signal separation in Ikeda and Murata, 1999; Parra and Spence, 1998; v. Gerven, 1996; Chan, 1997; Weinstein et al., 1993; Schobben and Sommen, 1998a. Output decorrelation uses only second order statistics which theoretically limits its separation capabilities as compared to higher order statistics. This is discussed in more detail in Section 6.2. Second order statistics do have the advantage that they can be estimated more reliably using less computational power than higher order statistics. The cross-correlations of the outputs can be expressed in terms of the cross-correlations of the observed signals and the known unmixing system, see Chan, 1997. These cross-correlations are non-linear functions of the unmixing filter coefficients and therefore the same holds for the cost-function that is composed of these cross-correlations. For stationary source signals, the cross-correlations of the observations can be estimated once and the cost function can be minimized using a gradient search. The problem is that thousands of coefficients need to be estimated in audio signal separation algorithms from a cost-function that is a non-linear function of these coefficients. This is hard to do with a gradient type of algorithm. Therefore, frequency domain approaches have been used recently in which the signal separation is done more or less independent per frequency, see Ikeda and Murata, 1999; Parra and Spence, 1998; Schobben and Sommen, 1998a. Details on such frequency domain approaches are given in Section 6.3. For non-stationary sources, the cross-correlations are updated so that the optimization algorithm must find a solution from a dynamically changing cost-function. The BSS algorithm that is presented in the next chapter is also based on output decorrelation.

Higher Order Statistics

The success of using higher order statistics in the instantaneous case Cardoso, 1989; Cardoso and Laheld, 1996; Amari et al., 1996 has boosted the interest in applying higher order statistics to the convolutive case over the last few years. Learning rules have been derived based on minimum mutual information, maximum entropy, divergence from Gaussianity, minimization of cross-HOS, Bussgang approaches and neural networks. These approaches are closely related and give learning rules that are very similar in most cases. Typically, non-linear elements are used that implicitly introduce HOS. The paper Bell and Sejnowski, 1995b is a good starting point for background, derivation and application of HOS in BSS. This work has been used later by other researches in convolutive BSS systems for audio applications. The next subsections describe instantaneous blind signal separation. Extensions for convolutive systems can be found in Lambert, 1996; Smaragdis, 1998 and Section 6.3. In Lee, 1998 background material is given for the techniques that are commonly used in this context.

Probability Density Functions

According to the central limit theorem, adding independent signals with a non-Gaussian PDF yields a signal that is closer to Gaussian than the separate independent signals. Therefore, when at most one of the sources has a Gaussian PDF, signal separation can be achieved by driving the recovered sources away from Gaussian PDFs. The Gaussianity can be measured directly from higher order statistics which are all zero for a Gaussian variable. Often, the kurtosis is used which is defined as $E\{y^4\} - 3(E\{y^2\})^2$. Large positive and negative values of the kurtosis correspond with peaky and flat PDFs respectively. Note that this technique requires prior knowledge on the direction to which the PDFs should be driven from Gaussian.

Another approach to measure the independence among the outputs is based on the factored marginal density of the outputs $\underline{y}^J = (y_1 \ldots y_J)^T$

$$p_{\underline{y}^J}(\underline{y}^J) = \prod_i p_{y_i}(y_i)\,. \tag{6.2}$$

This equation holds true iff the sources are independent. The independence of the outputs y_i can be expressed by the Kullback-Leibler divergence between the left and right side of (6.2)

$$\begin{aligned} D_{\mathrm{KL}}\left(p_{\underline{y}^J}(\underline{y}^J), \prod_i p_{y_i}(y_i)\right) &= \int p_{\underline{y}^J}(\underline{y}^J) \log\left(\frac{p_{\underline{y}^J}(\underline{y}^J)}{\prod_i p_{y_i}(y_i)}\right) d\underline{y}^J \\ &= -H(\underline{y}^J; W) + \sum_{i=1}^{J} H(y_i; W), \end{aligned} \tag{6.3}$$

with $H(y_i; W)$ the entropy of the marginal distribution p_{y_i} and $H(\underline{y}^J; W)$ the joint entropy of the outputs. Note that (6.3) is equal to the mutual information between the outputs. The difference among the approaches presented in the literature is on how to estimate $H(y_i; W)$ which cannot be calculated directly, see also for example Comon, 1994; Amari et al., 1996. An overview of prior knowledge that can be used for BSS is given in Jutten and Cardoso, 1995. One example is telecommunication where the probability density functions of the sources are often known in advance. Minimization of the mutual information is akin to information maximization which is described next.

Information Maximization

In information maximization the the outputs of the unmixing system $\underline{y}^J$ are passed through nonlinear elements that approximate the cumulative densities of the sources. If the true source signals are recovered at the outputs, the nonlinear elements produce signals with a uniform PDF which is the PDF with the largest entropy. Therefore, maximizing the entropy of the signals that are produced by the nonlinear elements corresponds to recovering the true sources. The learning rule derived in Bell and Sejnowski, 1995b is given by

$$W := W + \alpha \left((W^T)^{-1} - f(\underline{y}[n])(\underline{x}[n])^T \right),$$

with the scalar unmixing matrix

$$W = \begin{pmatrix} w_{11} & \dots & w_{1J} \\ \vdots & \ddots & \vdots \\ w_{J1} & \dots & w_{JJ} \end{pmatrix}$$

and $f(.)$ a non-linear function. The hyperbolic tangent is often used for this non-linearity. The optimal non-linearity depends on the (unknown) PDFs of the source signals. This update equation has been refined by Amari et al., 1996 who post-multiplied the update term by $W^T W$ which corresponds to an optimal rescaling of the entropy gradient

$$W := W + \alpha \left(W - f(\underline{y}[n])(\underline{y}[n])^T W \right). \tag{6.4}$$

This update equation has two important advantages as compared to the previous one; the computationally intensive matrix inversion is eliminated and the algorithm exhibits a good performance even for ill conditioned mixing matrices.

6.2. Indeterminacies

The separating criterions that are mentioned in the previous section do not incorporate any specific knowledge on the individual sources signals. The objective is merely to obtain separated signals. Consequently, instantaneous BSS algorithms can recover the sources up to a permutation and a scaling, i.e.

$$\underline{y}^J[n] = P^J D^J \underline{s}^J[n],$$

with P^J a permutation matrix which is obtained from the identity matrix by interchanging columns. Matrix D^J is a diagonal matrix which accounts for the scaling factors that cannot be determined. In practical applications, the scaling indeterminacy can be fixed by normalizing the recovered signals so that they have equal powers. In some applications such as hearing aids, the permutation itself is not a problem. It can be annoying though when the BSS algorithm swaps the sources during its operation. In applications with strong direct paths and little crosstalk, most BSS algorithms typically recover the signal at the ith output that is strongest at the ith microphone.

For convolutive BSS algorithms, the permutation indetermination is the same. The scaling indetermination however changes to a filtering indetermination. This means that the BSS algorithm has achieved its goal when filtered versions of the source signals are recovered. When no restrictions are made upon the unmixing filters, this will result in recovered signals that do not sound natural. Typically such signals have a flattened power spectral density so that they are observed as whitened or high-pass filtered versions of the source signals.

Besides these basic indeterminacies, there are some specific indeterminacies that are related to the source signals, the mixing system and the information that is used. For instantaneous mixing and unmixing, these can be summarized as Tong et al., 1991; Molgedey and Schuster, 1994

- Mutually uncorrelated sources $\underline{s}^J = (s_1 \ldots s_J)^T$ can be recovered from an instantaneous mixture $\underline{x}^J = H^J \underline{s}^J$ using second order statistics when the mixing matrix H^J is full rank and there is a $\tau > 0$ such that for all $i \neq j$

$$\frac{E\{s_i[n]s_i[n-\tau]\}}{E\{s_i^2\}} \neq \frac{E\{s_j[n]s_j[n-\tau]\}}{E\{s_j^2\}}.$$

- Mutually independent sources can be recovered from an instantaneous mixture using higher order statistics when the mixing matrix is full rank and for all $i \neq j$

$$\frac{E\{s_i^4\}}{E\{(s_i^2)^2\}} \neq \frac{E\{s_j^4\}}{E\{(s_j^2)^2\}}.$$

Note that independent sources are uncorrelated, while the opposite statement need not be true. When sources that are have short auto-correlation sequences are mixed instantaneously for example, SOS are insufficient to recover them from the mixture. When these signals have non-zero fourth order moments, HOS can be used. The fourth order moments of sources with Gaussian PDFs are equal to zero so that the sources cannot be recovered using HOS either in this case.

In Van Gerven and Van Compernolle, 1995 the uniqueness was studied of convolutive BSS of mixed stationary independent signals based on SOS. It was concluded that additional constraints are needed and that it is still not guaranteed that an adaptive algorithm converges to the desired solution. The cross power spectrum of the outputs can be written as

$$\Phi_{y_i y_j}(z) = \sum_{k=1}^{J} A_{ik}(z) A_{jk}^*(z) \Phi_{s_k}(z),$$

with $A_{ik}(z)$ the transfer function from source s_k to output y_i and $\Phi_{s_k}(z)$ the power spectrum of source s_k. From this expression it follows that the true and the 'permuted' solutions are not the only ones. Setting the cross power spectrum to zero for the case of two sources and two sensors for example gives

$$A_{11}(z) A_{21}^*(z) \Phi_{s_1}(z) = -A_{12}(z) A_{22}^*(z) \Phi_{s_2}(z). \tag{6.5}$$

True signal separation is achieved when this equation holds for all spectra $\Phi_{s_1}(z)$ and $\Phi_{s_2}(z)$. This is the case when crosstalk is eliminated, e.g. $A_{21}(z) = A_{12}(z) = 0$. This equation can also be fulfilled for particular choices of the unmixing filters that depend on the spectra of the source signals without achieving signal separation. This is called the "phantom" solution in Van Gerven and Van Compernolle, 1995. The uniqueness of SOS based convolutive BSS of non-stationary sources that are correlated in time is shown in Schobben and Sommen, 1999b. The derivation and proof is included in Appendix B. From this work, the following requirements are mentioned

- The sources must be correlated in time. For white noise sources, there is no unique solution. The proof in Tong et al., 1991 can be used here, as the convolutive case is a generalization of the instantaneous case. The BSS problem can be solved up to a rotation using SOS. In Delfosse and Loubaton, 1996 SOS are used to whiten the signals and the remaining rotation is solved using HOS. When the signals are correlated in time, the cross-correlations among the outputs can be minimized over a larger number of lags so that the set of solutions is reduced. In Chan, 1997 it is argued that separation results can be improved further by also using time-lags that are beyond the span of the adaptive filter, using more correlation lags than the number of unknowns.

- When FIR filters are used for signal separation the number of solutions is theoretically limited. This has been termed the 'FIR constraint' in Weinstein et al., 1993.

- The autocorrelations of the source must not be proportional to each other. It is clear from (6.5) that the cross power spectra can be divided out in that case so that the solution is not uniquely defined.

- The sources must be non-stationary. The set of solutions for the unmixing system that fulfills (6.5) for certain $\Phi_{s_1}(z)$ and $\Phi_{s_2}(z)$ which are measured over a time-interval includes the desired solution. This set of solutions is different for measurements of $\Phi_{s_1}(z)$ and $\Phi_{s_2}(z)$ over other time-intervals but it always contains the desired solution. When the source signals exhibit a short-term stationarity such as speech signals do, the desired solution is found from the intersection of various solution sets. In this way, solutions such as the phantom solution are excluded. In Torkkola, 1997 short-time dependencies in speech signals are removed first. The echoes with longer delays are left intact however and blind deconvolution is applied to learn the echo-removal filter from the whitened signals. This echo-removal filter is applied to the original signals effectively removing the echoes but leaving the speech signals intact.

6.3. Frequency Domain Approaches

Extensions of the instantaneous BSS algorithms have been developed for the convolutive case by several researchers. Some algorithms successfully succeeded in separating real world audio signals that were artificially mixed using short filters. The problem of separating real recorded mixtures is however far more complex due to the long acoustic transfer functions. The filters must be at least of the same length as the mixing system when separating two speakers from two microphone signals to achieve perfect signal separation according to Section 2.5. In practice shorter filters are sufficient. This is due to the fact that the transfer functions in the mixing system do not have heavy tails. The reverberation time of the room determines the product of the filter length over the sample frequency. A typical real-world case involving two speakers that are recorded using two microphones sampled at 16 kHz requires finding 4 filters of length 1024 to obtain good separation for a room with a reverberation time of about 100 ms. Using larger filters typically does not further improve the performance of the separation system as a larger set of filter coefficients need to be estimated. A straightforward gradient descent method that optimizes such a large number of parameters does not have good convergence properties unless a decomposition into relevant parameters is made.

Implementation in the frequency domain provides a solution. Neglecting leakage from one frequency bin to another, all harmonic components can be considered to be instantaneously mixed and can be unmixed instantaneously too, independent of each other. Several researchers have applied instantaneous BSS independently to the data at every frequency bin. The instantaneous BSS algorithm then works on complex data and the non-linear function in (6.4) must be chosen accordingly. Smaragdis, 1998 proposes to let the hyperbolic tangent work on the real and the imaginary part of the data separately. Although there is intuitively no reason to work with the real and imaginary parts of the data separately, it works well for practical applications. A small experiment that is not included in this book with plotting the 3-dimensional PDF of complex data of a certain frequency bin yields a rotation-symmetrical plot; the phase of the frequency domain data is uniformly distributed and the amplitude is generalized Gaussian distributed. Letting the non-linear element operate on the amplitude of these signals would therefore intuitively make more sense. Experiments with real-world data do not show significant improvements however.

The permutation and scaling problem is especially important when performing instantaneous BSS separately for all frequencies. Even when perfect separation is achieved for all frequencies, different permutations for various frequencies results in reconstructed signals that are still mixed. Methods that are presented in the literature to fix the **permutation problem** are

- *using the unmixing matrix of the previous frequency bin as an initial state of the unmixing matrix of the current bin* Smaragdis, 1997. If the unmixing matrices of the adjacent frequency bins are close, this procedure is likely to give the same permutation for all frequencies. To ensure that the unmixing matrices are close, the data is zero padded prior to frequency transformation, so that the spectra are interpolated;

- *assuming that the envelopes of the reconstructed signals are continuous as a function of frequency* Ikeda and Murata, 1998. First the envelopes of the separated signals are estimated per frequency bin by averaging the absolute values of these signals over time. Then the permutations are chosen for all frequency bins by maximizing the correlation among the envelopes for all frequencies;

- *restricting the unmixing filters to be linear* Schobben and Sommen, 1998a; Parra and Spence, 1998. In frequency domain approaches, the time domain filters are zero padded prior to transformation. This is required to perform linear convolutions by using circular convolutions in frequency domain. Hence the frequency domain filters have certain properties which must be maintained during adaptation of the filter coefficients in the frequency domain. When the filter is transformed back to the time domain, a zero padded real valued filter should be obtained. The permutation problem is introduced when instantaneous BSS is applied for all frequencies independently without looking to the time structure of the data and the filters. This is avoided by starting from a convolutive time domain approach and implementing it in the frequency domain.

The scaling problem becomes a filtering problem for convolutive unmixing. This means that the BSS algorithm tries to find arbitrarily filtered version of the sources. Methods that are presented in the literature to solve the **filtering problem** are

- *applying post-filters to the separated signals* Chan, 1997. Unlike the next two methods this is not specifically applicable for frequency domain approaches. Consider for example Figure 6.1 with $H_{11} = H_{22} = 1$ and using $w_{11} = w_{22} = 1$. When the solution $w_{12} = -H_{12}$ and $w_{21} = -H_{21}$ is found for example, the separated signals are filtered by $\frac{1}{1-w_{12}w_{21}}$ (the w_{ij} represent z-polynomials in the notation). The signals that are obtained by this post-processing are the sources the way they would have been recorded by the microphones in the absence of all other sources. Although such inverse filters are not stable in general, they can be made stable by mirroring the poles that are outside the unit circle into the unit circle. This gives perceptually good results as the human ear is not very sensitive to phase. However, these poles must be recalculated after each update of the unmixing system which is too computationally intensive for long filters to be done in real time;

- *inverse filtering by the unmixing filters after separation per frequency bin* Ikeda and Murata, 1999. When instantaneous BSS is applied per frequency bin, an unmixing matrix W_k^J is found for every bin k. Separation is obtained from $W_k^J \underline{X}_k^J$ with $\underline{X}_k^J = ((\underline{X}_1^M[kB])_k \dots (\underline{X}_J^M[kB])_k)^T$, i.e. a vector containing the kth frequency components of all microphone signals. Once these signals are separated they can be multiplied by $(W_k^J)^{-1}$ yielding contributions of the independent source signals to all microphones
$$\underline{X}_k^J(s_i) = (W_k^J)^{-1} \check{\Lambda}_i^J W_k^J \underline{X}_k^J,$$
where $(\check{\Lambda}_i^J)$ selects the ith component, i.e. $(\check{\Lambda}_i^J)_{ii} = 1$ and it contains zeros elsewhere. When the separation is successful, the $\underline{X}_k^J(s_i)$ is a vector containing the kth frequency component of the contribution from source s_i to all J microphones. When the permutation problem is solved, an entry of this vector can be joined with that of all other frequencies to obtain the corresponding separated signal. Note that, similar to the previous method, the contribution of a source to a microphone in the absence of all other sources is recovered rather than the source signal itself;

- *normalizing the unmixing filters W_k^J in the frequency domain* as in Schobben and Sommen, 1998a; Smaragdis, 1998. When the unmixing matrices are found per frequency bin by decomposing a cross-correlation matrix for example, this unmixing matrix will depend on the energy of the signal. As there is no knowledge of the energy distribution of the source signals, many algorithms recover signals that have equal energies for all frequencies. When no normalization is used, signals with a flat spectrum are produced in that case. Ideally, the unmixing filters preserve the amplitudes of the spectra and merely adjust the phase. This can be accomplished approximately by normalizing the unmixing matrices
$$W_k^J := \frac{W_k^J}{||W_k^J||}$$

for all frequencies k with $||.||$ an appropriate norm such as the l_2 norm.

6.4. Discussion

HOS are theoretically favorable as they provide more information about the independence of the signals. No perceptually superior performance is reported in the literature however for practical implementations of HOS based BSS algorithms as compared to that of algorithms based on SOS. An advantage of SOS over HOS is that they can be estimated faster and more accurately even in the presence of noise at a low computational complexity.

At least as important as the chosen approach is the care that is taken to implement it. Knowledge about acoustics, microphones and speech is required to obtain implementations that are feasible for commercial applications. More sophisticated algorithms can be obtained when using for example a speaker detector which is used to switch between sources and intelligently update SOS or HOS estimates. Real-time implementations are required to get a feeling for the performance of an algorithm and improvements that can be made.

Chapter 7

A BLIND SIGNAL SEPARATION ALGORITHM

This chapter addresses the problem of separating multiple speakers from mixtures of these that are obtained using multiple microphones in a room. A new blind signal separation algorithm is derived which is entirely based on second order statistics. The algorithm can run in off-line or online (adaptive) mode. One of the advantages of this algorithm is that no assumptions are made about the probability density functions or other properties of the signals. The blind signal separation algorithm has been tested using microphones that pickup speech signals of speakers that are talking simultaneously to give an indication of its performance for real-world data.

7.1. Introduction

Blind Signal Separation (BSS) deals with the problem of recovering independent signals using only observed mixtures of these. For acoustic applications, these observed mixtures are signals of multiple microphones. In this context a convolutive separation algorithm is used, i.e. the separation consists of employing Multi-Channel Finite Impulse Response (MC-FIR) filtering to these signals. Recently, several algorithms have been developed for convolutive BSS. Most of the algorithms make use of Higher Order Statistics (HOS). These HOS based algorithms contain non-linear elements that can be tuned to the data to obtain a good performance. This chapter presents a new BSS algorithm which is based only on SOS and does not require any parameters to be tuned. This work is based on Schobben and Sommen, 1998a.

The remainder of this chapter is outlined as follows. In Section 7.2 the optimization criterion that will be used in the derivation of the BSS algorithm is described. The optimization is done by minimizing the cross-correlations among the outputs of the multichannel separating filter. To achieve a computationally inexpensive algorithm with fast convergence, this criterion is transformed to the frequency domain in Section 7.3. The filter coefficients are calculated in the frequency domain such that the cross-correlations become equal to zero. At this point, no restrictions are imposed to ensure that the filter coefficients correspond to filters of a given length in the time domain. This is discussed in Section 7.4 and a method is suggested to remedy this problem. After modifying the unmixing filters using this method in the time domain, the cross-correlations among the outputs are no longer equal to zero. As there are constraints in both time and frequency domain an iterative method is proposed which operates in both domains. The key issue to obtain a good performance in terms of separation and convergence is to find a suitable adaptation in the frequency domain which leaves the time domain constraint intact as much as possible. This time-frequency domain compliance is discussed in Section 7.5. To prevent the algorithm from whitening the signals, a normalization must be applied. This is discussed in Section 7.6. The previously mentioned building blocks together form the new Convolutive Blind Signal Separation algorithm; CoBliSS, which is summarized in Section 7.7. CoBliSS has been tested on microphone signals from people that are speaking simultaneously in a room. The experimental results are presented in Section 7.9. This chapter concludes with deriving an estimate of the computational complexity of CoBliSS and investigating the feasibility of a real-time implementation.

7.2. Optimization Criterion

The MC-FIR separation filter will be controlled by an algorithm that minimizes cross-correlations among the outputs of this filter. The notation is in accordance with Figure 7.1 which depicts the cascaded mixing and unmixing system. The independent sources $s_1 \ldots s_J$ are convolved with the mixing system H to obtain the sensor signals $x_1 \ldots x_J$. Throughout, both the number of sources and the number of sensors are equal to J. Time indexes are not mentioned explicitly in all formulas. The transfer function from the lth input to the mth output of

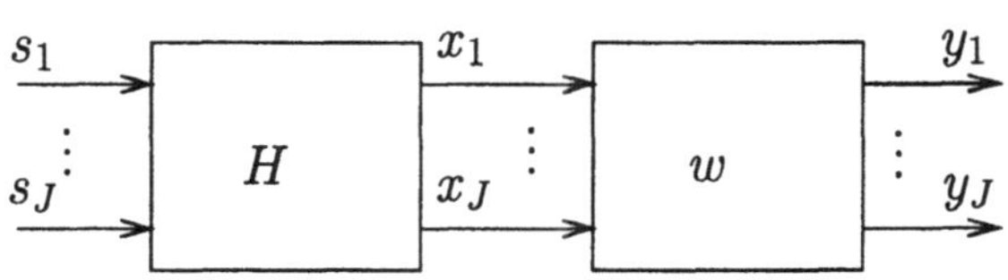

Figure 7.1. Cascaded mixing/unmixing system

the separation filter is denoted as $\underline{w}_{ml}^{N}$. The mth output of the separation filter

$y_m[n]$ is calculated from the observations $\underline{x}_l^N[n]$

$$y_m[n] = \sum_{l=1}^{J} (\underline{w}_{ml}^N)^T \underline{x}_l^N[n],$$

with J the number of microphones, N the filter length (all filters are of equal length for simplicity),

$$\underline{w}_{ml}^N = \begin{pmatrix} w_{ml}[N-1] \\ \vdots \\ w_{ml}[0] \end{pmatrix} \text{ and } \underline{x}_l^N[n] = \begin{pmatrix} x_l[n-N+1] \\ \vdots \\ x_l[n] \end{pmatrix}.$$

The cross-correlation among the outputs can be written as Chan, 1997

$$\begin{aligned} r_{y_i y_j}[l] &= E\{y_i[n] y_j[n-l]\} \\ &= E\left\{ \sum_{a=1}^{J} \sum_{b=0}^{N-1} w_{ia}[b] x_a[n-b] \sum_{c=1}^{J} \sum_{d=0}^{N-1} w_{jc}[d] x_c[n-l-d] \right\} \\ &= \sum_{a=1}^{J} \sum_{c=1}^{J} \sum_{b=0}^{N-1} \sum_{d=0}^{N-1} w_{ia}[b] w_{jc}[d] r_{x_a x_c}[l-b+d] \end{aligned} \tag{7.1}$$

with $r_{x_a x_c}[l] = E\{x_a[n] x_c[n-l]\}$. The advantage of this expression is that it can be used to optimize the filters w_{ia} using the cross-correlations of the observed data $r_{x_a x_c}$. These cross-correlations do not depend on the separation filters so that they do not need to be estimated again every time the separation filter is updated.

A cost function can be formed directly from (7.1) using for example the sum of squares of the cross-correlation coefficients. The straightforward minimization of such a non-linear cost function is not eligible however due to the large number of filter coefficients involved. A typical example is that two sources and two microphones are used. In that case 4 FIR filters need to be calculated, each with several hundreds or thousands of coefficients. All these coefficients highly depend on each other. Therefore, an approach is required that solves for filter coefficients subsets which are as independent of each other as possible.

This can be achieved by transforming (7.1) to the frequency domain. First, the cross-correlations are stacked in a vector for all correlation time-lags that are considered $l = l_1 \dots l_2$

$$\underline{r}_{y_i y_j}^L = \sum_{a=1}^{J} \sum_{c=1}^{J} R_{ac}^{L,2N-1} A_{jc}^{2N-1,N} \underline{w}_{ia}^N \tag{7.2}$$

with $L = l_2 - l_1 + 1$, and

$$\underline{r}_{y_i y_j}^L = \mathrm{J}^L \cdot \left(r_{y_i y_j}[l_1] \dots r_{y_i y_j}[l_2] \right)^T$$

$$A_{jc}^{2N-1,N} = \begin{pmatrix} w_{jc}[0] & 0 & \dots & 0 \\ \vdots & \ddots & \ddots & \vdots \\ \vdots & & \ddots & 0 \\ w_{jc}[N-1] & & & w_{jc}[0] \\ 0 & \ddots & & \vdots \\ \vdots & \ddots & \ddots & \vdots \\ 0 & \dots & 0 & w_{jc}[N-1] \end{pmatrix}$$

$$R_{ac}^{L,2N-1} = \begin{pmatrix} r_{x_a x_c}[l_2-N+1] & \dots & r_{x_a x_c}[l_2+N-1] \\ \vdots & & \vdots \\ r_{x_a x_c}[l_1-N+1] & \dots & r_{x_a x_c}[l_1+N-1] \end{pmatrix}.$$

Note that the cross-correlation matrix $R_{ac}^{L,2N-1}$ is identical to the Toeplitz cross-correlation matrix $(R_{ac}^{L})_{ij} = r_{x_a x_c}[j-i]$ for $l_2 = -l_1 = N-1$. It is desired for this matrix to have a Toeplitz structure as it can be approximated by a circulant matrix which in turn can be diagonalized in the frequency domain similar to the approach in Chapter 5. The advantage of rewriting the decorrelation criterion using diagonalized matrices is that the products of such matrices become element-wise vector products. The resulting expression can then be solved independently for all frequencies. The cross-correlation matrix $R_{ac}^{L,2N-1}$ can be extended to a square matrix of size $M \times M$ with $M \geq L$

$$R_{ac}^{L,2N-1} = \left(\mathrm{I}^{L} 0^{L,M-L}\right) R_{ac}^{M} \begin{pmatrix} 0^{l_2-N+1,2N-1} \\ \mathrm{I}^{2N-1} \\ 0^{M-N-l_2,2N-1} \end{pmatrix}$$

which merely requires that $l_2 \geq N-1$. This includes the unwindowed correlation matrix ($M = L = 2N-1$ and $l_2 = -l_1 = N-1$) as a special case. Substituting this extended cross-correlation matrix in (7.2) gives

$$\begin{aligned} \underline{r}_{y_i y_j}^{L} &= \sum_{a=1}^{J}\sum_{c=1}^{J} (\mathrm{I}^{L} 0^{L,M-L}) R_{ac}^{M} \begin{pmatrix} 0^{l_2-N+1,2N-1} \\ \mathrm{I}^{2N-1} \\ 0^{M-N-l_2,2N-1} \end{pmatrix} A_{jc}^{2N-1,N} \underline{w}_{ia}^{N} \\ &= (\mathrm{I}^{L} 0^{L,M-L}) \sum_{a=1}^{J}\sum_{c=1}^{J} R_{ac}^{M} \breve{A}_{jc}^{M} \begin{pmatrix} \underline{0}^{l_2-N+1} \\ \underline{w}_{ia}^{N} \\ \underline{0}^{M-l_2-1} \end{pmatrix} \end{aligned} \quad (7.3)$$

with $l_2 \geq N-1$, $M \geq N+l_2$. The circulant matrix $\breve{A}^M_{jc}$ is formed by extending $A^{2N-1,N}_{jc}$ on the right

$$\breve{A}^M_{jc}=\begin{pmatrix} w_{jc}[0] & 0 & \dots & 0 & \ddots & & \vdots \\ \vdots & \ddots & \ddots & & \ddots & \ddots & \vdots \\ \vdots & & \ddots & 0 & \dots & 0 & w_{jc}[N-1] \\ w_{jc}[N-1] & & & w_{jc}[0] & \ddots & & 0 \\ 0 & \ddots & & \vdots & \ddots & \ddots & \vdots \\ \vdots & \ddots & \ddots & \vdots & & \ddots & 0 \\ 0 & \dots & 0 & w_{jc}[N-1] & \dots & \dots & w_{jc}[0] \end{pmatrix}.$$

Note that in (7.3) it is used that for $l_2 \geq N-1$ and $M \geq N+l_2$

$$\begin{pmatrix} 0^{l_2-N+1,2N-1} \\ \mathrm{I}^{2N-1} \\ 0^{M-N-l_2,2N-1} \end{pmatrix} A^{2N-1,N}_{jc} = \breve{A}^M_{jc} \begin{pmatrix} \underline{0}^{l_2-N+1,N} \\ \mathrm{I}^N \\ \underline{0}^{M-l_2-1,N} \end{pmatrix}.$$

The cross correlation matrix R^M_{ac} can be approximated by its circulant variant $\breve{R}^M_{ac} = E\{(\breve{\mathcal{X}}_a[kB]^M)^T \breve{\mathcal{X}}_c[kB]^M\}$, with $\breve{\mathcal{X}}_l^M[kB]$ the circulant data matrix

$$\breve{\mathcal{X}}_l^M[kB]=\begin{pmatrix} x_l[kB-M+1] & x_l[kB] & \dots & x_l[kB-M+2] \\ \vdots & \ddots & \ddots & \vdots \\ \vdots & \ddots & \ddots & x_l[kB] \\ x_l[kB] & \dots & \dots & x_l[kB-M+1] \end{pmatrix}.$$

All elements of the time-domain cross-correlation expression are now rewritten such that they are eligible for transformation to the frequency domain in an efficient manner.

7.3. Frequency Domain Approach

In this section the cross-correlation expression in (7.3) is transformed to the frequency domain and a first step is made towards solving the BSS problem. Replacing R^M_{ac} by its circulant approximation in (7.3) makes it possible to diagonalize the matrices in this equation using FFTs, see Davis, 1979 and the previous chapters. This is done by inserting the identity matrix $(\mathcal{F}^M)^{-1}\mathcal{F}^M$ in between all matrices which results in

$$\underline{r}^L_{y_i y_j} = (\mathrm{I}^L\, 0^{L,L-M}) \sum_{a=1}^{J} \sum_{c=1}^{J} (\mathcal{F}^M)^{-1} \left(\underline{P}^M_{ac} \otimes \underline{W}^M_{jc} \otimes (\underline{W}^M_{ia})^* \otimes (\underline{V}^M)^{l_2} \right) \quad (7.4)$$

with

$$
\begin{aligned}
\underline{P}_{ac}^{M} &= \operatorname{diag}\left\{\mathcal{F}^{M}\breve{R}_{ac}^{M}(\mathcal{F}^{M})^{-1}\right\} \\
\underline{W}_{jc}^{M} &= \mathcal{F}^{M}\begin{pmatrix} \mathrm{J}^{N}\underline{w}_{jc}^{N} \\ \underline{0}^{M-N} \end{pmatrix} = \operatorname{diag}\left\{\mathcal{F}^{M}\breve{A}_{jc}^{M}(\mathcal{F}^{M})^{-1}\right\} \\
(\underline{W}_{ia}^{M})^{*} \otimes (\underline{V}^{M})^{l_2} &= \mathcal{F}^{M}\begin{pmatrix} \underline{0}^{l_2-N+1} \\ \underline{w}_{ia}^{N} \\ \underline{0}^{M-l_2-1} \end{pmatrix} \\
\underline{V}^{M} &= \begin{pmatrix} 1 & e^{\jmath\frac{2\pi}{M}} & \ldots & e^{\jmath\frac{2\pi(M-1)}{M}} \end{pmatrix}^{H}
\end{aligned}
$$

with J^{N} the $N \times N$ mirror matrix which has ones on its anti-diagonal and zeros elsewhere.

Signal separation is achieved when all cross-correlations among the outputs equal zero, i.e. $\forall i \neq j : \underline{r}_{y_i y_j}^{L} = \underline{0}^{L}$. Using as many correlation time-lags $L = M$ as possible for a given transform length M makes it possible to set all elements of (7.4) to zero directly in the frequency domain $\forall i \neq j$

$$
\sum_{a=1}^{J}\sum_{c=1}^{J} \underline{P}_{ac}^{M} \otimes \underline{W}_{jc}^{M} \otimes (\underline{W}_{ia}^{M})^{*} = \underline{0}^{M}. \tag{7.5}
$$

This expression has the advantage that the frequency domain filter coefficients can be calculated for all frequencies independently. As mentioned before, another advantage is that the cross-powers $\underline{P}_{ac}^{M}$ can readily be estimated from the microphone signals independent of the separation filters. The expression is reduced to a set of M equations with matrices of size $J \times J$. More precisely, there are two equations of real variables and $\frac{M}{2} - 1$ equations of complex variables due to the fact that both the filters and the correlation estimates are Fourier transforms of real vectors.

Next an approach is followed where all these equations are solved individually. The pth elements of $\underline{W}_{ij}^{M}$ and $\underline{P}_{ij}^{M}$ are put in a matrix $\forall i, j$

$$
W_p^J = \begin{pmatrix} (\underline{W}_{11}^{M})_p & \cdots & (\underline{W}_{1J}^{M})_p \\ \vdots & \ddots & \vdots \\ (\underline{W}_{J1}^{M})_p & \cdots & (\underline{W}_{JJ}^{M})_p \end{pmatrix} \tag{7.6}
$$

$$
P_p^J = \begin{pmatrix} (\underline{P}_{11}^{M})_p & \cdots & (\underline{P}_{1J}^{M})_p \\ \vdots & \ddots & \vdots \\ (\underline{P}_{J1}^{M})_p & \cdots & (\underline{P}_{JJ}^{M})_p \end{pmatrix}. \tag{7.7}
$$

In practical situations P_p^J is full rank[1], so that (7.5) can be rewritten $\forall p$ as

$$
(W_p^J)^{*} P_p^J (W_p^J)^{T} = \Lambda_p^J
$$

$$\Leftrightarrow (W_p^J)^T (\Lambda_p^J)^{-1} (W_p^J)^* \;=\; (P_p^J)^{-1} \tag{7.8}$$

with Λ_p^J a diagonal matrix. The off-diagonal zero elements in Λ_p^J are due to (7.5) meaning that the cross-correlations of the outputs of the BSS algorithm in frequency bin p must equal zero. The diagonal elements determine the autocorrelations of the outputs of the BSS algorithm in frequency bin p. Since Λ_p^J is real valued and its inverse is also diagonal, it can be absorbed by the weight matrices. For this reason, Λ_p^J is set equal to the identity matrix from now on. This implies that the autocorrelations of the outputs of the BSS algorithm in frequency bin p will equal 1. A normalization will be applied afterwards however on the unmixing filters to undo this as discussed in detail in Section 7.6. The P_p^J is symmetric by definition as the circulant cross-correlation matrices $\underline{P}_{ac}$ are symmetric, i.e. $\underline{P}_{ac} = \underline{P}_{ca}^*$. The inverse of this symmetric matrix P_p^J is also symmetric, so that the right hand side of (7.8) can be decomposed in many ways. In general the matrix decomposition should be different $\forall p$ in order to obtain signal separation. In practice these decompositions are unknown so that initially all P_p^J are decomposed in the same manner. When there is no prior knowledge on the microphone and speaker setup, it is feasible to start with a decomposition that is complex conjugate symmetrical, e.g. the matrix square root. This corresponds to a symmetric microphone and speaker setup. Therefore, the matrix square root will be used to obtain an initial guess for the weight matrices.

7.4. Convolution Constraint

Now that (7.8) is solved independently for all p using a matrix decomposition, the following no longer holds

$$(\mathcal{F}^M)^{-1} \underline{W}_{jc}^M = \begin{pmatrix} \underline{w}_{jc}^N \\ \underline{0}^{M-N} \end{pmatrix}. \tag{7.9}$$

In other words, the frequency domain filters no longer correspond to real-valued time domain filters of length N. The fact that the time domain filters must be real is not a problem, because the frequency domain cross correlation vectors have the same symmetric properties as the frequency domain filters. The fact that the time domain filters must be of length N is achieved by applying a time domain window $\forall j, c$

$$\underline{W}_{jc}^M := \mathcal{F}^M \begin{pmatrix} I^N & 0^{N,M-N} \\ 0^{M-N,N} & 0^{M-N} \end{pmatrix} (\mathcal{F}^M)^{-1} \underline{W}_{jc}^M. \tag{7.10}$$

This means that the filter coefficients that should be zero are set to zero in the time domain. Note that this operation strongly resembles that of the windowing in the frequency domain adaptive filters of the previous chapters. Clearly this modifies the weight matrices so that the solution of (7.8) no longer holds. There is a need for compliance between the frequency domain solution and the convolution constraint.

7.5. Time-Frequency Compliance

The two sets of equations (7.8) and (7.9) do not have a joint solution in closed form. Therefore an iterative approach is followed. The weight matrices are initialized once so that $W_p^T W_p^* = P_p^{-1}$. Then two steps must be performed which modify the weight matrices and (7.8) will no longer hold;

- the filters are constrained in the time domain according to (7.10).
- the cross-power matrices P_p are recursively estimated and these estimates vary slowly in time.

The key issue is to find a way to adapt the weight matrices slightly so that (7.8) holds again. This weight adaptation and (7.10) can be performed iteratively until convergence is achieved. This corresponds to finding the individual decompositions of P_p^J for all p as discussed in the previous section. Two methods are derived for the weight update; one is exact and one is an approximated version that exhibits a low computational complexity.

Exact Weight Update

In the following derivation, all matrices are of size $J \times J$ and the corresponding superscripts will be omitted. After constraining the filters according to (7.9) the weight matrix product becomes $W_p^T W_p^* = B_p$ with $B_p \neq P_p^{-1}$. Also, the cross-correlation matrices are updated $P_p \Rightarrow P_p'$. The goal is therefore to find a matrix C_p, such that $W_p'^T W_p'^* = P'^{-1}_p$ with $W_p' = W_p C_p$. The matrix C_p must be near to the matrix identity when B_p is near to $P_p'^{-1}$. In this way, the previous solution is preserved as much as possible and therefore fast convergence is ensured. Using the decomposed matrices

$$
\begin{aligned}
D_p = \text{sqrtm}\{B_p\}^* \quad &\Leftrightarrow \quad D_p^T D_p^* = B_p \\
D_p' = \text{sqrtm}\{P_p'^{-1}\}^* \quad &\Leftrightarrow \quad D_p'^T D_p'^* = P_p'^{-1}
\end{aligned}
$$

with sqrtm$\{.\}$ the matrix square root which is defined for a symmetrical matrix $B = B^H$ as $A = \text{sqrtm}\{B\}$ such that $A^H = A$ and $A^H A = B$. Note that $\text{sqrtm}\{B\}^* = \text{sqrtm}\{B^*\}$. The transform matrix C_p can be found from

$$
\begin{aligned}
B_p &= D_p^T (D_p'^T)^{-1} P_p'^{-1} (D_p'^*)^{-1} D_p^* \\
\Leftrightarrow W_p^T W_p^* &= D_p^T (D_p'^T)^{-1} W_p'^T W_p'^* (D_p'^*)^{-1} D_p^* \\
\Leftrightarrow \quad W_p &= W_p' (D_p')^{-1} D_p \\
\Leftrightarrow \quad W_p' &= W_p (D_p)^{-1} D_p'.
\end{aligned}
$$

So that

$$C_p = (D_p)^{-1} D'_p$$
$$= (\text{sqrtm}\{W_p^T W_p^*\}^*)^{-1} \text{sqrtm}\{P_p'^{-1}\}^*.$$

In an off-line implementation of the algorithm the cross-powers would first be estimated from all available data and $\text{sqrtm}\{P_p^{-1}\}^*$ would only have to be calculated once. In an online implementation however, the cross-correlations estimates change in time and require that the $\text{sqrtm}\{P_p^{-1}\}^*$ are recomputed every update. The matrix square root is computationally demanding when there are many signals to separate (large J). In the next subsection a method is derived with a low computational complexity that is more suitable for online implementation.

Fast Approximated Weight Update

A fast weight update is proposed in this subsection which does not use the computationally intensive matrix square root. Advantage is taken of the fact that the cross-power matrices change only slowly in time.

As in the previous subsection, $W_p^T W_p^*$ is no longer equal to P_p^{-1} after the time domain constraint is applied (7.10). Also the cross-power matrices are updated so that P_p changes to P'_p. Now a $J \times J$ matrix ϵ_p must be found such that

$$W_p'^T W_p'^* = P_p'^{-1} \quad \text{with} \quad W'_p = (\mathrm{I} + \epsilon_p) W_p. \tag{7.11}$$

So, new weight matrices must be derived from the previous ones so that their product becomes equal to the inverse of the updated cross-correlation matrices. Again, it is important that W'_p is close to W_p to achieve fast convergence of the algorithm. Accordingly it is desired that ϵ_p has a small norm. Using that P_p changes slowly in time and denoting $\Delta P_p = P'_p - P_p$ gives

$$W_p^T (\mathrm{I} + \epsilon_p)^T (\mathrm{I} + \epsilon_p)^* W_p^* = (P_p + \Delta P_p)^{-1}$$
$$W_p^T W_p^* + W_p^T (\epsilon_p^T + \epsilon_p^*) W_p^* \approx P_p^{-1} - P_p^{-1} (\Delta P_p) P_p^{-1}$$
$$\Leftrightarrow W_p^T (\epsilon_p^T + \epsilon_p^*) W_p^* \approx -P_p^{-1} (\Delta P_p) P_p^{-1}$$
$$\Leftrightarrow \epsilon_p^T + \epsilon_p^* \approx -(W_p^T)^{-1} W_p^T W_p^* (\Delta P_p) W_p^T W_p^* (W_p^*)^{-1}$$
$$\Leftrightarrow \epsilon_p^H + \epsilon_p \approx -W_p (\Delta P_p)^* W_p^H \tag{7.12}$$

The approximation corresponds to neglecting higher order terms of $P_p^{-1}(\Delta P_p)$ in the series expansion of $(\mathrm{I}+P_p^{-1}(\Delta P_p))^{-1}$. Also, it is assumed that $W_p^T W_p^* \approx P_p^{-1}$. This is only true when the weight matrices are not changed significantly by the time domain constraint (7.10). When the algorithm starts operating, both assumptions will be violated. Therefore, the exact weight update from the previous section may be used initially to get a good first estimate of the correlation and the weight matrices. Then the fast approximated weight update can be used. Now, ϵ_p must be chosen in accordance with (7.12) and such that the changes to W_p are small so that fast convergence is ensured. Both the left and right hand side of (7.12) are symmetric by definition. It follows from the triangle inequality that the ϵ_p with the smallest norm satisfying (7.12) is[2]

$$\epsilon_p = \epsilon_p^H = -\tfrac{1}{2} W_p(\Delta P_p)^* W_p^H.$$

In accordance with (7.11) the weight update becomes

$$\begin{aligned} W_p' &= (\mathrm{I} - \tfrac{1}{2} W_p(\Delta P_p)^* W_p^H) W_p \\ &= W_p(\mathrm{I} - \tfrac{1}{2}(\Delta P_p)^* W_p^H W_p) \end{aligned}$$

Note that in this notation W_p are the weight matrices that are readily constrained in the time domain as in (7.10).

7.6. Normalization

In Section 7.3 the constraint matrices Λ_p^J are set equal to the matrix identity. The impact of this is discussed in this section and a method to preserve the energy of the signals as a function of the frequency is given. The elements on the diagonal of Λ_p^J prescribe the power of the outputs of the separation filter at the corresponding frequency. Setting the constraint matrices Λ_p^J equal to the matrix identity is optimal for signals that have equal energy distributions as a function of frequency. Typically for real world signals like speech the energy decays significantly for higher frequencies. When the BSS algorithm is forced to yield outputs with equal energy for all frequencies this will result in energy boosting for frequencies where the signals are weak. Also, energy will be lowered for frequencies where the signals are strong. For speech for example, this leads to unwanted signal equalization where the low frequencies are suppressed and high frequencies are boosted resulting in artificially sounding recovered speech. This problem cannot be solved directly by choosing the Λ_p^J as they depend on the unknown original sources. Therefore the following approach is followed in which the W_p^J are calculated first from (7.8) with $\Lambda_p^J = \mathrm{I}^J$. Then the weight matrices are normalized using

$$W_p := \frac{W_p}{||W_p||}. \tag{7.13}$$

A norm that can be used and gives a good performance is the 2-norm which is defined as

$$||A^{K,L}|| = \sqrt{\sum_{i=1}^{K}\sum_{j=1}^{L}(A_{i,j}^{K,L})^2}\,.$$

The idea behind this is that all filter coefficients are of the same order of magnitude after this normalization is applied. Ideally all-pass filters are produced that leave the timbre of the signals unaffected.

7.7. Algorithm Overview

In this section the CoBliSS algorithm is presented which consists of the building blocks discussed in the previous sections. Note that operations such as matrix square root, matrix inverse and weight matrix normalization only need to be calculated for frequencies $p = 0, \frac{M}{2}$ (real valued matrices) and $p = 1 \ldots \frac{M}{2} - 1$ (complex matrices). The equations for higher frequencies are complex conjugated versions of these equations. The adaptive procedure consists of the following steps;

1 Blocks of microphone signals are transformed to the frequency domain for $a = 1 \ldots J$

$$\underline{X}_a^M = \mathcal{F}^M \begin{pmatrix} x_a[kB - M + 1] \\ \vdots \\ x_a[kB] \end{pmatrix}.$$

The blocks are of length M and are overlapping; only B new samples are used per block.

2 The cross-power estimates are updated efficiently in the frequency domain $\forall a, c : \underline{P}_{ac}^M := \alpha \underline{P}_{ac}^M + (1 - \alpha)\left((\underline{X}_a^M)^* \otimes \underline{X}_c^M\right).$
The forgetting factor α may vary from 0 to 1 depending on the application. Usually α is chosen near to 1, e.g. $\alpha = 0.99$ which is optimal for (nearly) stationary signals.

3 The weights are initialized by decomposing P_p^J (7.8) using the matrix square root
$\forall p : W_p^J = \text{sqrtm}\{(P_p^J)^{-1}\}^*$.
Note that $(P_p^J)_{a,c} = (\underline{P}_{ac}^M)_p$ as defined in (7.7).

4 The weights are modified such that the algorithm produces uncorrelated output in accordance with (7.8) (all matrices are of size $J \times J$)
$\forall p : W_p := W_p C_p$ with
$C_p = (\text{sqrtm}\{W_p^T W_p^*\}^*)^{-1} \text{sqrtm}\{P_p^{-1}\}^*$.
Note that this step can be omitted at initialization.

5 The weight matrices are normalized using the 2-norm
$W_p := \frac{W_p}{||W_p||}$.

6 The weights are constrained so that they correspond to linear Finite Impulse Response (FIR) filters of length N as in (7.9)

$$\underline{W}_{jc}^{M} := \mathcal{F}^{M} \begin{pmatrix} \mathrm{I}^{N} & 0^{N,M-N} \\ 0^{M-N,N} & 0^{M-N} \end{pmatrix} (\mathcal{F}^{M})^{-1} \underline{W}_{jc}^{M}.$$

Note that $(W_p^J)_{ac} = (\underline{W}_{ac}^M)_p$ as defined in (7.6).

7 The filtering is performed efficiently in the frequency domain using the overlap-save method to obtain the separated outputs

$$\underline{y}_j^B = (0^{B,M-B}\ \mathrm{I}^B)(\mathcal{F}^M)^{-1} \sum_{a=1}^{J} (\underline{X}_a^M \otimes \underline{W}_{ja}^M).$$

8 All steps are repeated iteratively except for the initialization in step 3.

Note that the filtering and the update can be calculated independently. Reducing the update rate lowers the computational complexity at the cost of a slower convergence.

7.8. Implementation Details

A few practical aspects must be taken into account when implementing the BSS algorithm. All microphone signals must be sampled synchronously. Problems can occur when the analog-to-digital converters that convert the microphone signals are not all controlled by the same clock signal. When one sample rate is slightly lower than the others, one signal will be delayed further and further compared to the others. Such data cannot be processed by a BSS algorithm. Another issue is that the digitized microphone signals must have mean values of zero. When the mean values are not zero, the correlation estimates are biased and the BSS algorithm will not work. Also, it is recommended that a good estimate of the cross-power vectors is made prior to initializing the weights and starting the adaptive procedure. If this is not done, the initialization is likely to be ill conditioned. A good estimate of the cross-power vectors can be obtained by performing step 2 of the algorithm about 2-10 times.

7.9. Experiments

The CoBliSS algorithm is implemented in Matlab. In this section, two experiments are presented with separating sources that are synthetically mixed. After verifying the operation of the algorithm on these signals, two experiments with real recorded data are presented. Finally, an experiment is done that is a first step towards joint blind signal separation and echo cancellation. In the next chapter the CoBliSS algorithm is generalized to efficiently deal with such cases.

It is difficult to give objective measures of quality of the separated signals. This is especially true for blind signal separation of real world recordings. Both the mixing system and the source signals are unknown in this case. The signals of three experiments in this section and the Matlab code of the CoBliSS algorithm are therefore included on compact disc for evaluation purposes. A description of the compact disc is given in Appendix E. In Chapter 9 it is described which cases are hard for signal separation. Also, methods for objective blind signal separation evaluation are derived which can be used in future research.

Separating Synthetic Mixtures

Initial experiments are done with instantaneously mixing two sinusoids $s_1[n]$ and $s_2[n]$ that have frequencies of 100 Hz and 105 Hz. The sample frequency is 44.1 kHz in this experiment. The mixing process is

$$\begin{pmatrix} x_1[n] \\ x_2[n] \end{pmatrix} = \begin{pmatrix} 0.7 & 0.3 \\ 0.3 & 0.7 \end{pmatrix} \cdot \begin{pmatrix} s_1[n] \\ s_2[n] \end{pmatrix} .$$

This mixing process is used to test how the convolutive blind signal separation algorithm operates on instantaneous mixtures. The mixing process does not reflect a real acoustic environment however. The filters that are used for signal separation have 128 taps. The number of new samples that is used each iteration of the block processing is 128. The forgetting factor in the cross-power estimation is $\alpha = 0.99$. It is chosen close to 1 as the source signals are stationary in this case. The source signals, the mixed signals and the recovered signals after processing with CoBliSS are depicted in Figure 7.2. It can be seen from this figure that the algorithm exhibits a fast convergence for the signals used. The long term behavior can be seen from Figure 7.3 where the same plots are depicted for $0.9 - 1.0$ s. The root mean squared (RMS) coefficients of the filters of the cascaded mixing and unmixing system are calculated. The ratio of the RMS of the direct paths over that of the cross-paths is only about 9 dB in this experiment. The remaining 'crosstalk' is inaudible however. This is because the unmixing filters only have to be good for the frequencies that are occupied by the source signals. Calculating the ratio of the power of the desired signal over the crosstalk readily gives about 27 dB after 1 second of processing.

An additional interesting experiment is to repeat the same test with independent noise added to both source signals. The noise power equals the power of the sinusoids in this case. The separation results are perceptually as good as before. The coefficient power ratio changes to about 23 dB and the signal power ratio remains unchanged. This is because the unmixing system can now be found over the whole frequency range as the excitation signals cover the whole frequency band. Comparing signal powers is therefore a good measure of quality compared to looking at the cascaded mixing and unmixing system.

Next a more difficult experiment is done with mixing two music signals using real measured acoustic impulse responses of 512 taps. FIR filters are used for signal separation which are also of length 512. This experiment is complicated in two ways; the source signals are not stationary and the mixing system is

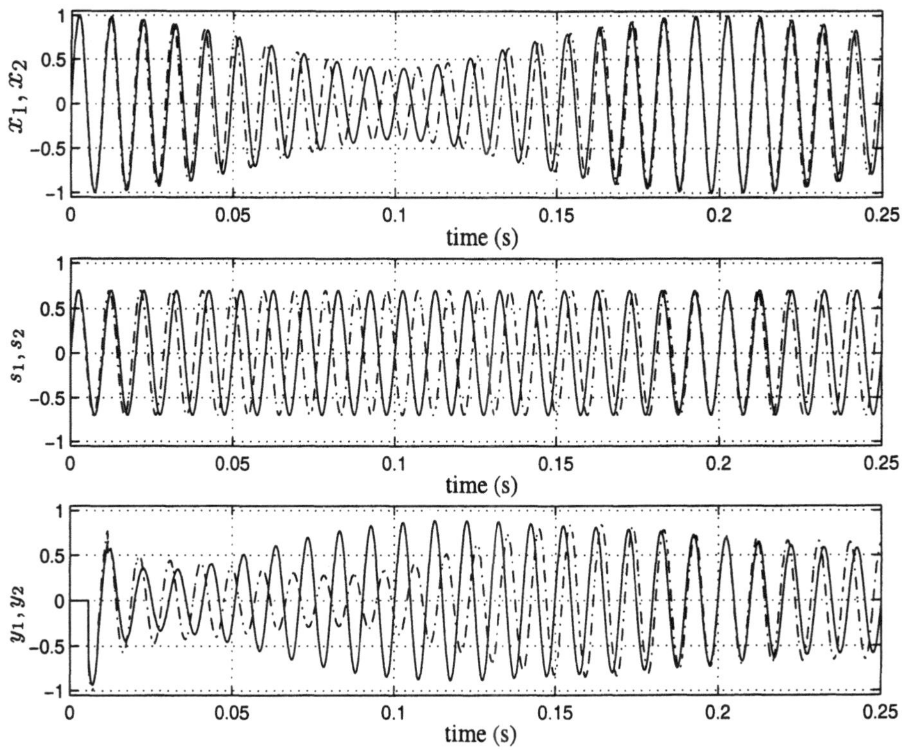

Figure 7.2. The instantaneously mixed signals x_1 and x_2, the source signals s_1 and s_2 and the recovered signals, 0.0-0.25 seconds

more complex than that of the previous experiment. The sample frequency is $f_s = 22050$ Hz and the forgetting factor in the cross-power update is $\alpha = 0.95$. The separation results are perceptually very reasonable and are included on compact disc. There is some remaining crosstalk but this is almost completely masked by the desired signals. This result gives no guarantees however for the performance of an online implementation of CoBliSS that operates on real recorded data. In experiments with real world data, the assumption that the mixing filters are of a certain finite length is violated. Also, there is additional microphone noise.

Separating Real World Mixtures

Experiments are done with audio recorded in a real acoustic environment. The room which is used for the recordings was 3.4 x 3.8 x 5.2 m (height x width x depth) and is depicted in Figure 7.4. Two male speakers simultaneously read 4 sentences aloud. In an initial experiment, the loudspeaker as depicted in Figure 7.4 is not used. The speakers are live speakers, not loudspeakers. The

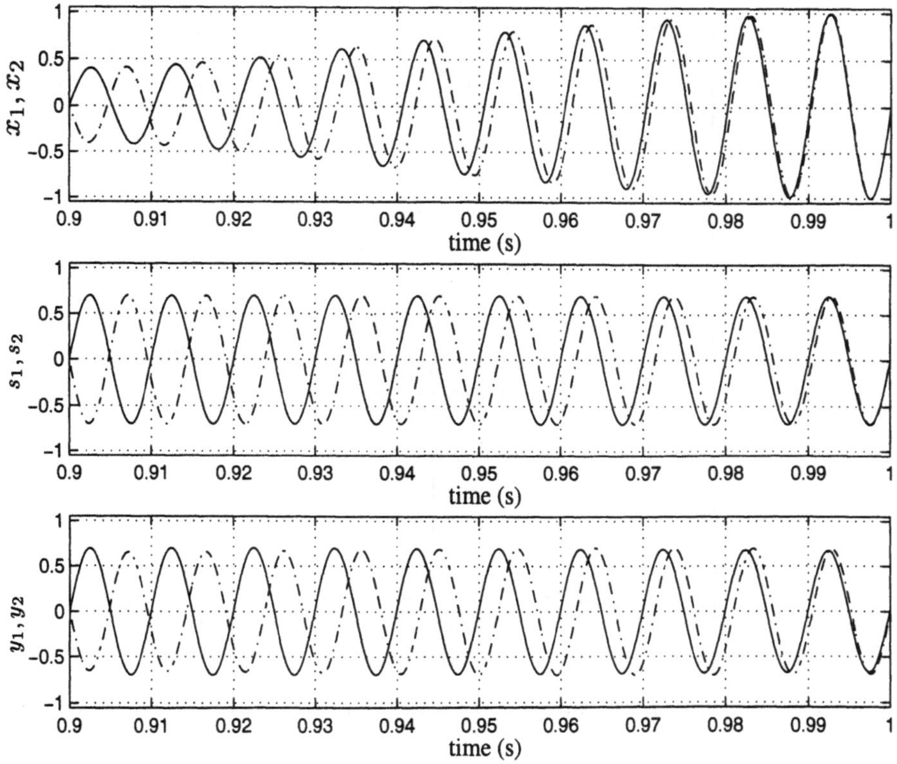

Figure 7.3. The instantaneously mixed signals x_1 and x_2, the source signals s_1 and s_2 and the recovered signals, 0.9-1.0 seconds

resulting signals are recorded by two microphones which are spaced 58 cm apart. The recordings are of 16 bit precision, sampled at 24 kHz and are 10 seconds long. The separation filters are of length 512 and are controlled by the CoBliSS algorithm using Fourier transforms of length 1024. The forgetting factor in the cross-power estimation is $\alpha = 0.95$. For the sake of computational complexity, only one update per every 2560 samples is done. In this experiment, the algorithm converges to a good solution within 0.25 second. This is verified by using the unmixing filters that are found at that time as fixed filters for unmixing the whole sound track. Further adaptation beyond 0.25 second gives signals that perceptually are slightly better separated. The microphone signals and the separated outputs are included on compact disc. There is still some crosstalk present in the separated signals although this is at a level which is no longer disturbing when listening to either of the recovered source signals. The timbre of the separated signals is almost the same as that of the recorded signals. Experiments have been done with longer filter lengths but no significant improvement is found for this case. Increasing the FFT size while keeping the

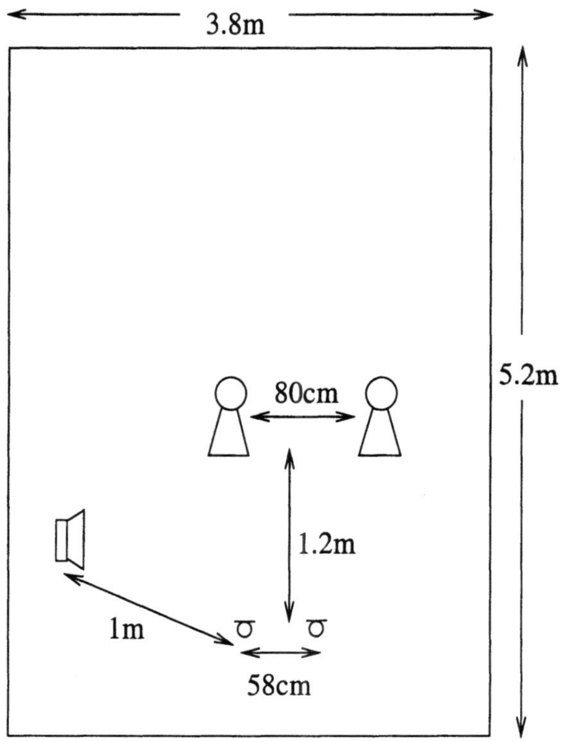

Figure 7.4. Recording setup

filter length constant gives slightly different separated outputs. The differences are small however so that no clear improvement can be reported.

Towards Combined BSS and AEC

The previous experiment is repeated with also playing French radio news over a small loudspeaker as depicted in Figure 7.4. The far-end speech is used as a third input for CoBliSS. Hence, the far-end speech signal is treated as any other microphone signal and the knowledge that is readily is an independent component is not exploited here. Applying CoBliSS to these three signals results in successful separation in the sense that one source signal is clearly stronger than the others in each output. Some crosstalk remains however, and the level of the crosstalk is higher than in the previous experiment in which only 2 speakers were present. The microphone signals, the far-end signal and the separated outputs are included on compact disc. The power of the crosstalk that is introduced in the 'recovered' far-end speech is less than the power of the crosstalk that remains in the recovered speakers. In conclusion CoBliSS leaves the independent far-end speech more or less unchanged and it aims at eliminating it from the other channels. This operation can be seen as a combined signal separation of two speech signals and the echo cancellation of a third signal in the two microphone signals. It has an important advantage though; the echo cancellation is no longer troubled by double talk. Double talk is the classical problem that the performance of an AEC is deteriorated when besides the far-end speech a local speaker is active too. It can even lead to instability of an AEC. When a BSS also performs AEC as in this experiment, it even is an advantage when all speakers and loudspeakers are active at the same time. When this is not the case, correlations can be estimated over a longer time to include contributions of all speakers or a (loud)speaker activity detector can be used in estimating the correlations or in switching between the speakers. Such implementation aspects are very important for the behavior of CoBliSS in practical situations.

Clearly, CoBliSS can be extended using the knowledge that the far-end speech readily is a source signal as is shown in Schobben and Sommen, 1998c; Schobben and Sommen, 1999a. In this way, the quality of the recovered sources can be further improved and the computational complexity can be reduced significantly. This extended CoBliSS is derived in the next chapter.

7.10. Computational Complexity

In this section, the feasibility of a real-time implementation of the CoBliSS algorithm is investigated. An efficient implementation is proposed exploiting the properties of the matrices involved and the corresponding computational complexity is discussed. The work in this section and Appendix C is based on Schobben and Sommen, 1998b.

It is assumed that a complex multiplication is performed using 4 real multiplications and 2 real additions. As a measure of complexity, the number of floating point operations is used which is assumed to be equal to the sum of real multiplications and additions. An overview of the total complexity that is required to compute one recursion of the CoBliSS algorithm is given in Table 7.1. The fast approximated weight update from Subsection 7.5 is used. The derivation of the computational complexity of the entries in this table is included in Appendix C. The item numbers in the table correspond to the steps of the CoBliSS algorithm in Section 7.7. In a typical numerical example the filters are of length $N = 1024$, the FFT size is $M = 2048$, $B = 1024$ new samples are used each block, the sample rate is $f_s = 44.1$ kHz, and $J = 2$ microphones are used. The computational complexity evaluated for these parameters amounts to 29 Mega-flops per second. The complexity is also evaluated as a function of the number of microphones J for different transform sizes M as shown in Figure 7.5. The sample frequency equals 44.1 kHz for this figure. Note that the complexity of the algorithm depends linearly on the sample frequency. A typical personal computer achieves 1000 Mega-flops per second. Therefore a real-time implementation of CoBliSS is feasible, even for more than 2 microphones.

Table 7.1. Computational Complexity Overview

Operation	Item	Complexity (flops)
(I)FFTs	1. 6. 7.	$(J^2 + J) \cdot 5M \log_2(B)$
Correlation update	2.	$3J^2M$
Weight update	4.	$\frac{M}{2}(12J^3 - 13J^2 + 5J)$
Normalization	5.	$\frac{M}{2}5J^2$
Filtering	7.	$\frac{M}{2}(2J^2 - J)$

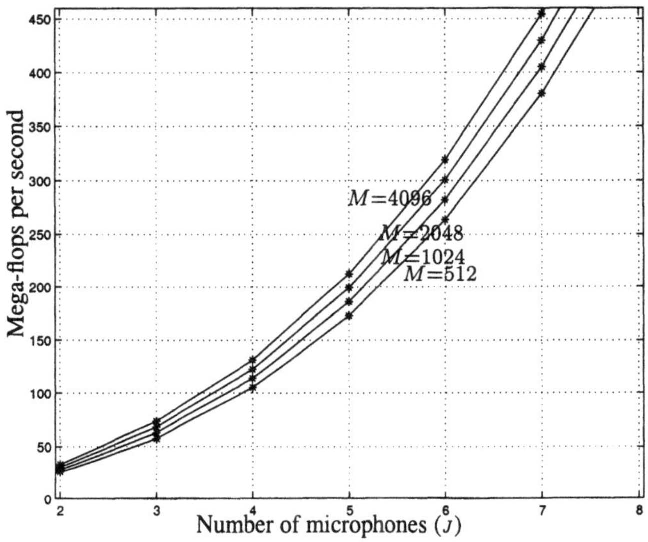

Figure 7.5. Overall complexity of the CoBliSS algorithm as a function of the number of microphones J and the transform size M, with $f_s = 44.1$kHz

Figure 7.6 shows both the overall complexity (upper curves) and the part of this complexity that is due to the FFTs (lower curves). Increasing transform sizes ($M = 512, 1024, 2048, 4096$) correspond to higher complexity curves similar to Figure 7.5. If can be seen from this figure that the FFTs take most of that computational complexity when less than 10 microphones are used. According to table 7.1 the complexity of these FFTs depends quadratically on the number of microphones. Therefore, the overall complexity approximately depends quadratically on the number of microphones. For more than 10 microphones, the CoBliSS weight update becomes computationally intensive. The complexity is than proportional to J^3 and real-time operation is no longer feasible.

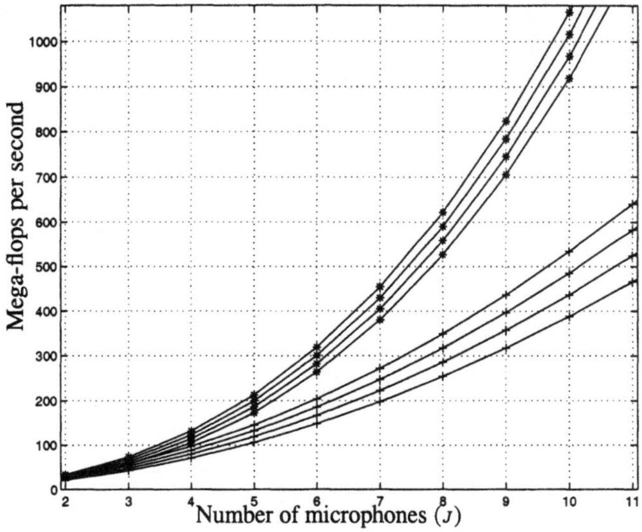

Figure 7.6. Overall complexity ('*') and the complexity due to the FFTs ('+') of the CoBliSS algorithm as a function of the number of microphones J and the transform size M, with $f_s = 44.1$kHz

7.11. Conclusions

A new blind signal separation algorithm is presented which is based on second order statistics. First experiments with separating real world recordings show that the algorithm has a very acceptable performance when applied to real world audio problems involving speech and music. The computational complexity of the algorithm quadratically depends on the number of microphones. Real-time implementation is feasible on a personal computer when using up to 10 microphones.

Notes

1 Typically, the microphone signals have energies of the same order of magnitude for a frequency. This can be enforced if necessary by normalizing the microphone signals for each frequency. Matrix P_p^J is properly conditioned as it has a relatively strong diagonal as the autocorrelations are stronger than the cross-correlations. When there are fewer sources than J, P_p^J will become ill conditioned. Also, if the sources do not occupy the whole frequency range this matrix will be ill conditioned. A practical solution for this is to use regularization; if the condition number of P_p^J is below a certain threshold, $P_p^J := P_p^J + \epsilon \mathrm{I}^J$ with ϵ some small number. Note that microphone noise also induces such kind of regularization.

2 This can be seen by writing ϵ_p as a symmetric and a skew-symmetric part and observing that the skew-symmetric part does not contribute to $\epsilon_p + \epsilon_p^H$. The ϵ_p with the smallest norm is therefore symmetrical.

Chapter 8

JOINT BLIND SIGNAL SEPARATION AND ECHO CANCELLATION

The problem of joint blind signal separation and acoustic echo cancellation arises in applications such as teleconferencing and voice-controlled machinery. Typically, in the same room as the local speakers there are loudspeakers reproducing far-end speech or music signals. The contributions of these loudspeaker signals to the microphone signals need to be canceled. The remaining signals are then separated so that the individual local speakers are recovered. In the previous chapter the CoBliSS blind signal separation algorithm is introduced. In this chapter an extension of CoBliSS is presented (ECoBliSS) which can simultaneously deal with blind signal separation and echo cancellation. The computational complexity of ECoBliSS is less than that of CoBliSS when processing the same number of signals. The performance of the extended CoBliSS algorithm is evaluated using audio recorded in a real acoustic environment. This chapter is based on Schobben and Sommen, 1999a.

8.1. Introduction

In an application such as voice-controlled television, a user should be able to control the television while other people are having a conversation nearby. Typically the television has stereophonic sound and microphones are used to pick up the commands of the user. These microphone signals contain a mixture of speaker signals and television loudspeaker signals. A straightforward approach would be to use multichannel AEC as presented in Chapter 5 followed by BSS as presented in the previous chapter. This is depicted in Figure 8.1. This approach has the following two important drawbacks however

- the AEC does not work well when there is double talk. This means that it can only operate reliably when none of the speakers is talking. When people walk through the room and therefore influence the acoustic transfer functions

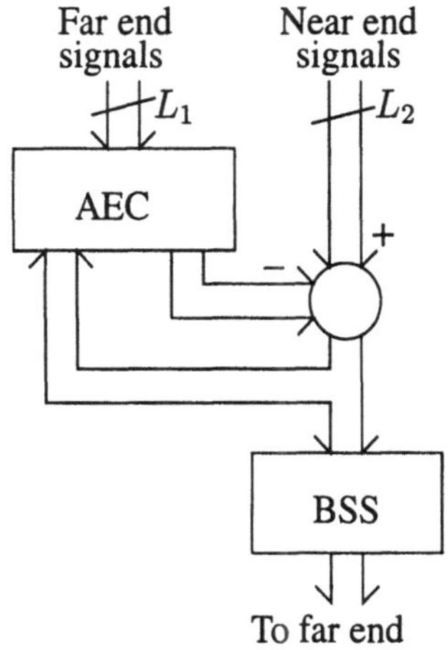

Figure 8.1. Echo cancellation followed by blind signal separation

from the loudspeakers to the microphones fast adaptation is required which cannot be obtained in the presence of double talk.

- The BSS algorithm is hampered by contributions of the loudspeaker signals that remain present in the microphone signal despite of the AEC. The BSS can no longer find independent signals and the overall system performance deteriorates accordingly.

Applying BSS to both the microphone signals and the far-end signals would overcome these drawbacks but it is not feasible due to the limited performance of blind systems and their high computational complexity. Also the permutation of the output signals is not known. It is therefore desirable to extend CoBliSS accordingly such that it can perform simultaneous BSS and AEC.

The problem of recovering source signals from mixtures of them which are contaminated by acoustic echoes is depicted schematically in Figure 8.2. The $s_1 \ldots s_{L_2}$ are the unknown sources, e.g. the local speakers. These sources are assumed to be independent of each other. The $s_{L_2+1} \ldots s_J$ are the known sources, e.g. far-end signals that are reproduced in the same room. These far-end signals typically are not independent of each other as for example in a system that produces stereophonic sound. They are assumed to be independent of the local speakers though. The multichannel room impulse response is modeled by H and contains all transfer functions from the speakers and the loudspeakers to the microphones. The microphone signals are denoted as $x_1 \ldots x_{L_2}$. The correlation estimator measures the cross-correlations among all microphone signals and the known sources. This information is used to update the multichannel filter w which produces $y_1 \ldots y_{L_2}$ as outputs. These outputs are the estimates of the unknown sources $s_1 \ldots s_{L_2}$. The multichannel filter

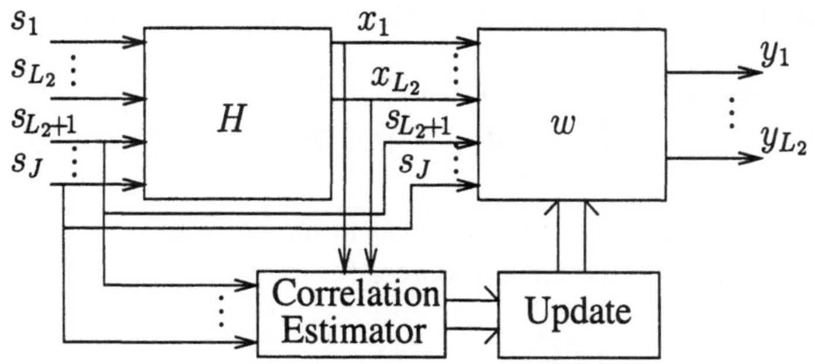

Figure 8.2. Mixing, unmixing and control system

has FIR filters between each of its J inputs consisting of both the microphone signals and the far-end signals and its L_2 outputs.

8.2. Derivation

For notational convenience the far-end signals are assigned to $x_j = s_j$ for $j = L_2 + 1 \dots J$ so that the x_j now represent the input signals for ECoBliSS for $j = 1 \dots J$. The filter part can now be written for $m = 1 \dots L_2$

$$y_m[n] = \sum_{l=1}^{J} (\underline{w}_{ml}^N[n])^T \underline{x}_l^N[n]$$

with $\underline{x}_l^N[n] = (x_l[n-N+1] \dots x_l[n])^T$. The $\underline{w}_{ml}^N[n]$ perform blind signal separation for $l = 1 \dots L_2$ and acoustic echo cancellation for $l = L_2 + 1 \dots J$. The idea that forms the basis of ECoBliSS is to make all output signals mutually uncorrelated and uncorrelated with the far-end signals.

Similar to CoBliSS, the time domain input signals are transformed to the frequency domain and the cross-correlation estimates are updated in the frequency domain. For every frequency, the filter coefficients and the correlation estimates are grouped similar to (7.7)

$$W_p^{L_2,J} = \begin{pmatrix} (\underline{W}_{11}^M)_p & \cdots & (\underline{W}_{1J}^M)_p \\ \vdots & \ddots & \vdots \\ (\underline{W}_{L_21}^M)_p & \cdots & (\underline{W}_{L_2J}^M)_p \end{pmatrix} = \begin{pmatrix} \dot{W}_p^{L_2} \, \ddot{W}_p^{L_2,L_1} \end{pmatrix}$$

$$P_p^J = \begin{pmatrix} (\underline{P}_{11}^M)_p & \cdots & (\underline{P}_{1J}^M)_p \\ \vdots & \ddots & \vdots \\ (\underline{P}_{J1}^M)_p & \cdots & (\underline{P}_{JJ}^M)_p \end{pmatrix}.$$

The $\dot{W}_p^{L_2}$ correspond to the filters that perform the BSS and the $\ddot{W}_p^{L_2,L_1}$ correspond to the filters that perform the AEC. The decorrelation criterion that is written as $(W_p^J)^* P_p^J (W_p^J)^T = \Lambda_p^J$ in the previous chapter now changes to

$$\begin{pmatrix} (\dot{W}_p^{L_2})^* & (\ddot{W}_p^{L_2,L_1})^* \\ 0^{L_1,L_2} & \mathrm{I}^{L_1} \end{pmatrix} P_p^J \begin{pmatrix} (\dot{W}_p^{L_2})^T & 0^{L_2,L_1} \\ (\ddot{W}_p^{L_2,L_1})^T & \mathrm{I}^{L_1} \end{pmatrix} = \begin{pmatrix} \Lambda_p^{L_2} & 0^{L_2,L_1} \\ 0^{L_1,L_2} & \breve{P}_p^{L_1} \end{pmatrix} \quad (8.1)$$

with $\Lambda_p^{L_2}$ a diagonal constraint matrix as in the previous chapter and $\breve{P}_p^{L_1}$ the cross-power matrix of the known sources for frequency bin p

$$\breve{P}_p^{L_1} = \begin{pmatrix} (P_{L_2+1,L_2+1})_p & \cdots & (P_{J,L_2+1})_p \\ \vdots & \ddots & \vdots \\ (P_{J,L_2+1})_p & \cdots & (P_{J,J})_p \end{pmatrix}. \quad (8.2)$$

The weight matrix in (8.1) consists of the matrices $\dot{W}_p^{L_2}$ and $\ddot{W}_p^{L_2,L_1}$, a zero matrix and a identity matrix. This can be seen as a filter structure that yields both $y_1 \ldots y_{L_2}$ and $s_{L_2+1} \ldots s_J$ as outputs. The right hand side of (8.1) prescribes that $y_1 \ldots y_{L_2}$ are uncorrelated as Λ_p^J is a diagonal matrix. The zero matrices prescribe that the $y_1 \ldots y_{L_2}$ are not correlated with the far-end signals $s_{L_2+1} \ldots s_J$. The $\breve{P}_p^{L_1}$ is required as the cross-correlations among the far-end signals remain unchanged. Equation (8.1) is rearranged as

$$\begin{pmatrix} (\dot{W}_p^{L_2})^T & 0^{L_2,L_1} \\ (\ddot{W}_p^{L_2,L_1})^T & \mathrm{I}^{L_1} \end{pmatrix} \begin{pmatrix} (\Lambda_p^{L_2})^{-1} & 0^{L_2,L_1} \\ 0^{L_1,L_2} & (\breve{P}_p^{L_1})^{-1} \end{pmatrix} \begin{pmatrix} (\dot{W}_p^{L_2})^* & (\ddot{W}_p^{L_2,L_1})^* \\ 0^{L_1,L_2} & \mathrm{I}^{L_1} \end{pmatrix} = (P_p^J)^{-1} \quad (8.3)$$

so that

$$\begin{pmatrix} (\dot{W}_p^{L_2})^T (\Lambda_p^{L_2})^{-1} (\dot{W}_p^{L_2})^* & (\dot{W}_p^{L_2})^T (\Lambda_p^{L_2})^{-1} (\ddot{W}_p^{L_2,L_1})^* \\ (\ddot{W}_p^{L_2,L_1})^T (\Lambda_p^{L_2})^{-1} (\dot{W}_p^{L_2})^* & (\ddot{W}_p^{L_2,L_1})^T (\Lambda_p^{L_2})^{-1} (\ddot{W}_p^{L_2,L_1})^* \end{pmatrix} = (P_p^J)^{-1} - \begin{pmatrix} 0^{L_2} & 0^{L_2,L_1} \\ 0^{L_1,L_2} & (\breve{P}_p^{L_1})^{-1} \end{pmatrix}. \quad (8.4)$$

Note that both sides of this equation are of rank L_2 (with P_p^J and $\breve{P}_p^{L_1}$ full rank)[1] so the weight matrix products are uniquely defined. The inverse of the correlation matrix is now written as

$$(P_p^J)^{-1} = \begin{pmatrix} \dot{P}_p^{L_2} & \ddot{P}_p^{L_2,L_1} \\ \dot{\breve{P}}_p^{L_1,L_2} & \ddot{\breve{P}}_p^{L_1} \end{pmatrix}.$$

So that the two relevant equations from (8.4) are

$$(\dot{W}_p^{L_2})^T (\Lambda_p^{L_2})^{-1} (\dot{W}_p^{L_2})^* = \dot{P}_p^{L_2} \quad (8.5)$$

$$(\dot{W}_p^{L_2})^T (\Lambda_p^{L_2})^{-1} (\ddot{W}_p^{L_2,L_1})^* = \ddot{P}_p^{L_2,L_1}. \quad (8.6)$$

Again, the diagonal matrix $(\Lambda_p^{L_2})^{-1}$ can be absorbed by the weight matrices. Therefore, the BSS matrix $\dot{W}_p^{L_2}$ can be initialized by using the matrix square root to decompose the inverted cross-correlation matrix in the same way as in CoBliSS

$$\dot{W}_p^{L_2} = (\text{sqrtm}\{(\dot{P}_p^{L_2})^{-1}\})^*.$$

The AEC matrix $\ddot{W}_p^{L_2,L_1}$ depends on the BSS matrix $\dot{W}_p^{L_2}$. It is found from rewriting (8.6)

$$\ddot{W}_p^{L_2,L_1} = (\dot{W}_p^{L_2})^{-H}(\ddot{P}_p^{L_2,L_1})^*. \quad (8.7)$$

In this approach the AEC weight matrix is calculated from the BSS weight matrix. This is an advantage over traditional systems with separate BSS and AEC, where the BSS is updated independent of the AEC. When the AEC changes in such a system, the BSS has to re-converge.

When the weights are initialized, the cross-correlation estimates are updated and the weights are constrained so that they correspond to linear FIR filters of length N as in (7.9). Equations (8.5) and (8.6) no longer hold after these modifications. Similar to CoBliSS, the $\dot{W}_p^{L_2}$ are post-multiplied by $C_p^{L_2}$ to correct this, with

$$C_p^{L_2} = (\text{sqrtm}\{(\dot{W}_p^{L_2})^T(\dot{W}_p^{L_2})^*\}^*)^{-1}\text{sqrtm}\{(\dot{P}_p^{L_2})^{-1}\}^*.$$

The AEC matrix can again be found from (8.7). This procedure is repeated continuously. Together, the previous steps form the ECoBliSS algorithm which is summarized in the next section.

8.3. Algorithm Overview

Although ECoBliSS strongly resembles CoBliSS there are several differences. The ECoBliSS algorithm consists of the following steps;

1 Blocks of input data are transformed to the frequency domain for $a = 1 \dots J$

$$\underline{X}_a^M[kB] = \mathcal{F}^M \begin{pmatrix} x_a[kB-M+1] \\ \vdots \\ x_a[kB] \end{pmatrix}.$$

These blocks are of length M and are overlapping; only B new samples are used per block.

2 Cross-power estimates are updated efficiently in the frequency domain $\forall a, c$:

$$\underline{P}_{ac}^M[kB] \;\; := \;\; \alpha\underline{P}_{ac}^M[(n-1)B] + (1-\alpha)\left((\underline{X}_a^M)^*[kB] \otimes \underline{X}_c^M[kB]\right).$$

The forgetting factor α may vary from 0 to 1 depending on the application. Usually α is chosen near to 1, e.g. $\alpha = 0.99$. In contrast to CoBliSS, these cross-powers are calculated among both the microphone signals and the far-end signals.

3 When the cross-correlation matrices are initially estimated the weights that correspond to the signal separation are initialized using the matrix square root
$\forall p : \dot{W}_p^{L_2} = \text{sqrtm}\{\dot{P}_p^{L_2}\}^*$.
The weights that correspond to the acoustic echo cancellation are calculated
$\forall p : \ddot{W}_p^{L_2,L_1} = ((\dot{W}_p^{L_2})^H)^{-1}(\ddot{P}_p^{L_2,L_1})^*$.

4 The weights are changed such that (8.1) holds again
$\forall p : \dot{W}_p^{L_2} := \dot{W}_p^{L_2} C_p^{L_2}$
with $C_p^{L_2} = (\text{sqrtm}\{(\dot{W}_p^{L_2})^H \dot{W}_p^{L_2}\})^{-1}\text{sqrtm}\{\dot{P}_p^{L_2}\}^*$.
and $\forall p : \ddot{W}_p^{L_2,L_1} = ((\dot{W}_p^{L_2})^H)^{-1}(\ddot{P}_p^{L_2,L_1})^*$.
Note: This step can be omitted if the initialization has just been done.

5 The weight matrices are normalized using the 2-norm
$\dot{W}_p^{L_2} := \frac{\dot{W}_p^{L_2}}{\left\|\left(\dot{W}_p^{L_2} \quad \ddot{W}_p^{L_2,L_1}\right)\right\|}$ and $\ddot{W}_p^{L_2,L_1} := \frac{\ddot{W}_p^{L_2,L_1}}{\left\|\left(\dot{W}_p^{L_2} \quad \ddot{W}_p^{L_2,L_1}\right)\right\|}$.
Note that $\left(\dot{W}_p^{L_2} \quad \ddot{W}_p^{L_2,L_1}\right)$ means that $\dot{W}_p^{L_2}$ and $\ddot{W}_p^{L_2,L_1}$ are both placed side-by-side in one matrix.

6 The weights are constrained so that the frequency domain weights corresponds to time domain filters of length N, $\forall p$:
$$\underline{W}_{jc}^M := \mathcal{F}^M \begin{pmatrix} \mathrm{I}^N & 0^{N,M-N} \\ 0^{M-N,N} & 0^{M-N} \end{pmatrix} (\mathcal{F}^M)^{-1} \underline{W}_{jc}^M.$$
Note that $(W_p^J)_{ac} = (\underline{W}_{ac}^M)_p$.

7 The filtering is performed efficiently in the frequency domain using the overlap-save method to ideally obtain the separated and echo free outputs
$\underline{y}_j^B = (0^{B,M-B} \ \mathrm{I}^B)(\mathcal{F}^M)^{-1} \sum_{a=1}^J (\underline{X}_a^M \otimes \underline{W}_{ja}^M)$.

8 All steps are repeated iteratively except for the initialization in step 3.

Note that similar to CoBliSS, the filtering and the weight update can be calculated independently. When the update of the cross-correlations is slower than the weight update for example, reducing the weight update rate lowers the computational complexity at the cost of only a slightly slower convergence of the system.

8.4. Experiments

The last experiment of Chapter 7 where CoBliSS was used on two microphone signals and a loudspeaker signal is repeated here using ECoBliSS. Again, the room that is used for the recordings is depicted in Figure 7.4. Two persons read

4 sentences aloud. Far-end speech was introduced by playing the French news over a small loudspeaker. The resulting sound was recorded by two microphones which were spaced 58 cm apart. The recordings are sampled at 24 kHz, with 16 bit accuracy. The separation filters are of length 512 taps. Extended CoBliSS only calculates the two separated speaker signals. The outputs of extended CoBliSS sound significantly better than those of the first experiment. Also, the computational complexity of extended CoBliSS is significantly lower, as the number of filters decreases and the matrices that need to be decomposed become smaller.

The signals that are recovered by ECoBliSS in this experiment are included on compact disc.

8.5. Conclusions

An extension of the CoBliSS BSS algorithm is presented in this chapter. This extension is capable of performing joint acoustic echo cancellation and blind signal separation at a low computational complexity. The performance is verified using data that is collected in a real world environment. The extended algorithm exhibits a performance that is superior to that of the CoBliSS algorithm when the known signal is used as an additional input. The computational complexity of the extended CoBliSS algorithm is significantly lower which makes it suitable for real-time operation. An additional advantage of the CoBliSS algorithm over conventional echo cancellation is that it can operate in double talk situations without complications. This makes it suitable for applications such as teleconferencing and hands free telephony.

Notes

1 For the left hand side of (8.4) this is obvious because it equals the left hand side of (8.3) with $(\breve{P}_p^{L_1})^{-1}$ replaced by zeros. For the right hand side it can be seen from

$$\text{rank}\left\{\begin{pmatrix} A & B \\ C & D \end{pmatrix}^{-1} - \begin{pmatrix} 0 & 0 \\ 0 & D^{-1} \end{pmatrix}\right\} =$$

$$\text{rank}\left\{\text{I} - \begin{pmatrix} A & B \\ C & D \end{pmatrix}\begin{pmatrix} 0 & 0 \\ 0 & D^{-1} \end{pmatrix}\right\} = \text{rank}\left\{\begin{pmatrix} \text{I} & -BD^{-1} \\ 0 & 0 \end{pmatrix}\right\}.$$

Chapter 9

BLIND SIGNAL SEPARATION ALGORITHM EVALUATION

Recently, many new Blind Signal Separation (BSS) algorithms have been introduced. Authors evaluate the performance of their algorithms in various ways. Among these are speech recognition rates, plots of separated signals, plots of cascaded mixing/unmixing impulse responses and signal to noise ratios. Clearly, not all of these methods give a good reflection of the performance of these algorithms. Moreover, since the evaluation is done using different measures and different data, results cannot be compared. As a solution a unified methodology of evaluating BSS algorithms is provided along with data that is available such that researches can compare their results. Although this chapter is focussed on acoustic applications, many of the remarks apply to other BSS application areas as well. The work in this chapter is published in Schobben et al., 1999[1].

9.1. Introduction

It seems that BSS can have a large number of applications in the audio realm, especially in the area of signal enhancement by removing undesired source components from a desired signal, see Girolami, 1998; Lee et al., 1997b; Sahlin and Broman, 1997; Schobben and Sommen, 1998c; Yen and Zhao, 1996. Thus, this area has recently received a lot of attention. However, currently it is not possible to compare different algorithms reliably as every research paper seems to measure a different aspect of the performance using a different dataset.

The purpose of this chapter is to remedy this problem by discussing what is needed for BSS performance evaluation. Obviously there is no single perfect measure of goodness since there is no single definition of the problem (ICA, separation, deconvolution). In addition to merely evaluating the success of a separation algorithm, a more ambitious problem is to construct a set of tests of variable difficulty examining a set of distinct properties that would provide valuable information about the weaknesses of the algorithm. This implies full control over the separation problem which in turn implies the need for synthetic test cases. Synthetic cases can be used to examine algorithm performance in trivial up to ill-conditioned cases thus rating separation ability accurately. By having control over the type of the sources, the effect that they might have on algorithm performance can be seen.

However, synthetic cases fail to capture some elements of the real world. There is a certain level of complexity in the real world which cannot confidently be reproduced by synthetic means. The statistics of room responses, the dynamic quality of the convolution (even at seemingly static cases), factors such as the physical presence of the sources in the environment and the particular patterns of background noise, are hard to reproduce but present additional complications worth examining.

Thus a suite of test cases is proposed which is divided into two main categories:

1 A number of controllable synthetic separation problems are provided. They test the limits of algorithms covering a wide range of attributes. As the sources are available performance measurement is straightforward.

2 With real world recordings clean sources are not available and measuring the separation quality is often difficult. A test methodology is provided that combines the best of both worlds: the realism of true audio recordings in real environments, and the ability to accurately measure the separation performance on these recordings. This comes from recording each source *separately* in a real environment as described later.

Providing methodologies along with data sets makes it possible for researchers to compare their algorithms in a more objective way.

The rest of this chapter is organized as follows. First will be discussed what aspects of BSS tasks could be measured and controlled to characterize BSS algorithms. A few most important ones are chosen for synthetic test cases. Next, possible measures of performance are discussed together with the recording setup that is used for the evaluation of BSS systems. This is important since the recording setup determines what data is available for the evaluation of the BSS algorithm. A setup is described that combines realism with easy performance measurement. Finally the chosen test cases are discussed which are included on the CD-rom.

9.2. On the Difficulty of Blind Signal Separation

Before starting to discuss measures that indicate the degree of signal separation achieved, conditions that could increase the difficulty of a BSS task are considered. These conditions are thus candidates for parameters to vary when constructing the test cases.

Convolutive mixing of the sources is inherent in almost all imaginable audio BSS applications. In addition, some aspects are enumerated that are related to instantaneous mixing. As the whole chapter, the enumerated conditions are geared towards audio situations:

1 The closer the mixing is to a singular matrix the harder the separation task is for algorithms that do not exhibit the equivariant behavior, see Amari et al., 1996; Cardoso and Laheld, 1996. In the presence of noise the task becomes harder also for equivariant algorithms. The level of difficulty can be controlled by adjusting the eigenvalue spread of the mixing filter matrix, see Lambert, 1996.

2 There is a continuum from instantaneous mixing to delayed mixing, i.e. convolutive mixing with only one nonzero coefficient per filter. This can be used to measure the ability of an algorithm to deal with simple convolutive mixing.

3 There is also a continuum from delayed mixing to real world convolutive mixing, which can be explored by changing the sparseness and the duration of the mixing filters. This, tested, can rate an algorithm's ability to deal with increasingly complicated mixing filters.

   In real recordings these aspects can be controlled to some extent by changing the positions of the microphones and sources. The easiest cases are in general when the mixing matrix has strong direct paths with little crosstalk; i.e. every source is close to its microphone. Also the acoustic characteristics of the recording room can be controlled (anechoic vs. hard walled chamber). Introducing more reverberation makes the separation task more difficult in general.

4 In any kind of a mixing situation the probability density functions (PDF) of the sources have an effect. Usually the closer they are to Gaussians, the harder the separation gets. This is because many BSS algorithms maximize the distance of the recovered source to Gaussianity.

5 The spectra of the sources may vary from narrowband to wideband, which can have great influence on the performance of the algorithm. Tests should include sounds of both classes since some algorithms might rely on these qualities.

6 Some algorithms make use of the *difference* of the spectra of different source signals. Therefore it is useful to include test cases with distinct source spectra and test cases with similar source spectra.

7 Also in any kind of mixing the available amount of data needed to successfully learn to separate a *static* mixing situation characterizes how well the

algorithm might perform in *dynamic* mixing circumstances. There is a continuum from static mixing to rapidly varying mixing. This can be used to vary the level of difficulty when testing an algorithm's tracking capabilities. When there is no comprehensive data set available with dynamic mixtures tracking capabilities can be judged from the convergence of the algorithm on static mixtures.

8 The ability to deal with silences is also needed, at least for static algorithms. Sections of silence from a source should not cause the algorithm to diverge. For example in the presence of background noise an algorithm should not give a wildly different estimate of the separation system when the speakers are silent for short periods of time. The system would then have to reconverge each time the speakers appear again.

9 Increasing the number of sources together with the number of mixtures increases the degree of difficulty significantly. For example, algorithms that work well in the 2-by-2 case might fail miserably in the 4-by-4 case. At the limit of convolved unmixing there is a 1-by-1 case which corresponds to blind deconvolution.

10 Keeping the number of sources fixed but varying the number of available mixtures can greatly influence the behavior of the algorithm. In general, at least the same number of mixtures as the number of sources is required. If there is further information available lesser number might suffice. By using more mixtures than there are sources, the capabilities of the algorithm to tolerate noise or to improve the separation performance could be characterized.

11 The amount and the quality of noise in the mixtures can be controlled using:

(a) A single noise signal independent of all sources mixed to each sensor signal.

(b) Different noise components, independent of all sources and each other, mixed to each sensor signal.

(c) Similarity of the noise PDF/spectrum to the source signals.

Together, if all these aspects could be characterized, a fairly complete picture of the capabilities of an algorithm could be obtained. Synthetic test cases that cover these aspects are introduced and discussed in Section 9.4.

9.3. Recording Setups and Performance Measures

In this section it is discussed how to evaluate the goodness of the actual separation on the basis of test data. In addition to separation, BSS algorithms can be characterized by distortion, i.e., how the original signals are distorted from how each microphone would observe them in the absence of other sources.

It appears that there are three fundamentally different methods to evaluate separation depending on what data is available.

The first one is based on the impulse responses of the mixing channels and the separation filters. By convolving the mixing and separation systems, it is simple to measure how far (in dB) the resulting filters are from a scaled unit response. As the mixing channels are required this method only applies to the synthetic case. Although this method is well defined it might not accurately evaluate the separation of the signals. For example, when the sources have low energy contents for certain frequency intervals the BSS algorithm might fail in finding an unmixing system that achieves separation for those frequencies. This does not affect the quality of separation as there is almost no frequency content in the signals for these frequencies.

The second one is based on the test signals themselves. For each separated signal the residuals of the other sources are compared against the desired source. Ideally these residuals should equal zero. Note that to be able to do this for a real recording, static mixtures are required, in which only one source is active at a time, together with a labeling that indicates these locations in the test mixtures.

A third way is to directly evaluate the independence of the separated outputs. It is difficult however to come up with a measure of independence that can be estimated accurately and gives a clear indication of quality of separation.

In the following subsection a recording setup is described that enables evaluation of real recordings using the second method mentioned above and overcoming its limitations.

Recording Setup

The source signals $s_1 \dots s_J$ are filtered by the multichannel acoustic transfer function H yielding the microphone signals $x_1 \dots x_J$ as in Figure 7.1. It is assumed that the number of sources equals the number of microphones. For synthetic cases the source signals are filtered by the premeasured multichannel acoustic transfer function H. In that case the unmixing system w can be evaluated using the known mixing system H and the unmixed signals y_i can be judged using the known source signals s_i. This makes it possible to evaluate distortion and separation.

When real recordings are used however, the only available data are the microphone signals x_i, $i = 1 \dots J$. Both the separation and the distortion that is introduced by the unmixing system cannot be determined directly from the microphone signals.

The following approach combines the realism of true audio recordings in real environments, and the ability to accurately measure the separation performance on these recordings. In this approach multichannel recordings are made in a room when only one of the sources is active at a time. These recorded signals are denoted x_{i,s_j}, i.e. the contribution of the jth source to the ith microphone. The mixed data is obtained by adding these independent contributions for all sources, i.e. $x_i = \sum_j x_{i,s_j}$. Now the x_i represent the microphone signals that would have been obtained if all sources were active simultaneously. This is justified for acoustic applications, since sound waves are additive too.

Note that all speakers must be present during the recordings, even if they are silent, as the room acoustics are influenced by their presence. The approach is therefore limited to the recording of non-moving sources, as the source movements cannot be reproduced exactly in general. Using this approach, multichannel recordings are as realistic as they can be without losing information that is required for the evaluation. A method for measuring the quality of separation and measuring the distortion due to the BSS algorithm using these recorded signals is described in the next subsection.

Performance Measures

Distortion

The distortion of the jth separated output can be defined as

$$D_j = 10 \log \left(\frac{E\{(x_{j,s_j} - \alpha_j y_j)^2\}}{E\{(x_{j,s_j})^2\}} \right)$$

with $\alpha_j = E\{x_{j,s_j}^2\}/E\{y_j^2\}$. The separated signal indices are chosen such that y_j corresponds to the jth source. In this definition, the separation results are not distorted when they are equal (up to a scaling factor α_j) to x_{j,s_j}, i.e. the contribution to microphone j of the jth source alone. The permutation and scaling of the signals do not affect the distortion and are therefore left out of the definition.

When synthetic mixtures are used this approach can be followed too as the x_{i,s_j} can be calculated from the mixing system and the sources s_j so that the same approach can be followed.

A more detailed impression of the distortion can be obtained by plotting it as a function of frequency using a Short Time Fourier Transform (STFT)

$$D_j(\omega) = 10 \log \left(\frac{E\{|\text{STFT}\{x_{j,s_j} - \alpha_j y_j\}|^2\}}{E\{(x_{j,s_j})^2\}} \right).$$

Problems arise however when convolutive signal separation is considered. The α_j then must compensate for both gain and propagation delay which is hard to estimate accurately. A more favorable approach is to compare

$$E\{|\text{STFT}\{x_{j,s_j}\}|^2\}/E\{(x_{j,s_j})^2\} \text{ vs. } E\{|\text{STFT}\{y_j\}|^2\}/E\{y_j^2\}$$

in one plot. From such plot it can be seen if and how the timbre of the signals is affected. Matlab code to generate such plots is included on the CD-rom.

The baseline for distortion is the original microphone signal. Whatever happens to the signal from the source to the microphone cannot be determined as the actual sources are not available. This distortion measure is thus biased to favor methods that do not perform deconvolution in addition to separation as any deconvolution would be observed as distortion.

Separation

The quality of separation of the jth separated output can be defined as

$$S_j = 10\log\left(\frac{E\{(y_{j,s_j})^2\}}{E\{(\sum_{i\neq j} y_{j,s_i})^2\}}\right) \tag{9.1}$$

with y_{j,s_i} the jth output of the cascaded mixing/unmixing system when only s_i is active. Other definitions of quality of separation involving the mixing/unmixing system are less suited since BSS is about signal separation and not about system identification. The separation can also be plotted as a function of the frequency using

$$S_j(\omega) = 10\log\left(\frac{E\{(\text{STFT}\{y_{j,s_j}\})^2\}}{E\{(\text{STFT}\{\sum_{i\neq j} y_{j,s_i}\})^2\}}\right). \tag{9.2}$$

Matlab code for the evaluation of separation is also included on the CD-rom.

9.4. Data Sets

Synthetic Tests

Data is included on the CD-rom which can be used for the evaluation of BSS algorithms. This suite includes synthetic data and non-synthetic data. Both subsets contain the same original sources, mixed in synthetic and real environments respectively.

Source Signals

The set of sources includes various speech phrases, music passages, environmental sounds and synthetic tones.

The speech sources consist of the same set of sentences read by different speakers. The sentences include various ratios of vowel/consonants so as to have some elementary indication of bandwidth. Given that several sentences are read by the same speakers, this can be used to provide mixtures of these, on which heuristic assumptions will be harder because of source similarity.

The music sources consist of a set of recordings of various characteristics. Music passages are included of various spectral shapes to measure the dependency of an algorithm on these characteristics of sound, as well as its behavior with respect to the relation of these characteristics across sources i.e. a narrowband and a wideband source, a high and a low centroid source etc. There are also music passages featuring wide dynamic changes to test how well an algorithm can track sources that suddenly fade out or even disappear.

Environmental sounds include common types of background noise, mostly to provide an indication of performance for the case of speech/background noise mixing. The examples include street noise as a fullband signal which can completely cover the bandwidth of speech and 'hide' it very effectively. Certain sounds which because of their sparse and self similar character are hard to eliminate are also included (e.g. bouncing ball). Also a few relatively narrowband machinery sounds are included which exhibit a variety of spectrums.

Finally the synthetic sounds provide basic tests on the influence of bandwidth, PDFs, frequency variance and interruptions. The set consists of a sine wave, a square wave, a sawtooth wave, Gaussian noise and Cauchy noise. Where applicable the frequency of the sound can change so as not to provide a completely stationary source. The same waveforms are also provided with random interruptions during their progress to test how well an algorithm can track disappearing sources. The evaluation of these cases will be easier since plots of the outputs can be provided and intuitive observations on the performance of the algorithms can be made out of these.

Mixing

For the synthetic tests the above sources were mixed in arbitrary groups using a set of different mixing situations, including instantaneous mixing matrices, sparse convolutive mixing matrices, dense convolutive mixing matrices and estimates of real world mixing matrices.

The set of instantaneous mixing matrices includes both well and ill conditioned problems. In addition some cases are contaminated with added noise in the sources. This set can measure the speed and quality of convergence as well as the accuracy of an algorithm, under both clean and degenerate conditions. The order of the matrices is varied to test how well an algorithm can deal with an arbitrary number of sources.

The set of sparse convolutive matrices includes a few mixing matrices spanning the range between simple delayed mixing to more complicated filters derived out of theoretical analysis of room reflections. This test will rate how algorithms perform with convolutive mixtures of increasing complexity.

For the dense convolutive mixing matrices, the set contains cases of dense convolutive matrices whose filters are obtained from various kinds of manipulated noise sequences. Real room impulse responses are very similar to exponentially decaying noise with exponential or Cauchy distribution, so such series are used as filters. The parameters of the series comprising the delays are tuned to various levels to control complexity. Fast decay and high kurtosis are the simplest cases since they produce short and sparse filters, whereas longer decay times and lower kurtosis will generate harder cases. An additional degree of complexity is also provided where at the simplest level there is only one filter in the mixing matrix which is dense, while the rest are just delay taps. The number of dense filters in the mixing matrix can be progressively increased to cover all cases and reach the hardest one where all filters are dense.

Finally there are two cases which use filters measured from a real room and a dummy head used for binaural recordings. These are real, dense mixing matrices which can come as close as synthetically possible to the real world. They are used to get an accurate reading of performance (since the mixing matrix is known) before moving to the real cases.

Real Recordings as Test Cases

The real world recordings are done in two different rooms; a near anechoic room and a typical living room. Live speakers are used and their contributions are recorded independently. Music sources are reproduced using loudspeakers. The same music tracks are used as in the synthetic case. Again, all of these contributions are recorded independently so that the performance measures from Subsection 9.3 can be applied. As the music signals are known, algorithms that solve BSS together with for example acoustic echo cancellation can be evaluated, too. The clean speech recordings that are done in the near anechoic room are also used as source signals for the synthetic case. The transfer functions in the near anechoic room resemble a simple delayed mixing matrix which should be relatively easy to deal with for most BSS algorithms. The living room, however, corresponds to a mixing matrix with dense filters. Estimates of the room mixing matrices can be estimated from the loudspeaker and the microphone signals too so that they also can be used for the synthetic cases. Also, a dummy head is used for binaural recordings. This data can be used to evaluate BSS algorithms for applications like hearing devices where the microphones are relatively small and cheap and shadowed by the head.

Evaluation Software

The following Matlab files have been included on the CD-rom

- A plot routine which plots the power spectral density of the source contribution to a microphone and the recovered source as a function of frequency
- An evaluation routine for the quality of Blind Signal Separation as described in (9.1)
- A plot routine which plots the separation as a function of frequency as described in (9.2)
- A utility that reads multiple wav-files that contain a multi-track recording

Notes

1 This paper is co-authored by Kari Torkkola, Motorola, Phoenix Corporate Research Labs, Tempe, USA and Paris Smaragdis, MIT Media Lab, Cambridge, USA

Chapter 10

CONCLUSIONS

10.1. Main Contributions

This book is focussed on multichannel signal processing concepts for audio applications. Two new multichannel signal processing approaches are presented in this book being a frequency domain multichannel echo canceler and a frequency domain blind signal separation algorithm based on second order statistics. In addition, efficient implementations are presented for the single channel echo canceler and the delay and sum beamformer. Also, methods to evaluate blind signal separation algorithms are given along with comments on what cases are difficult for blind signal separation algorithms.

The multichannel echo canceler is a frequency domain implementation of the multichannel recursive least squares algorithm. The bottleneck in (multichannel) RLS is the computationally intensive inversion of the (multichannel) cross-correlation matrix. A well-known efficient frequency domain implementation of single-channel RLS is generalized in this book for multichannel RLS. In this approach, the multichannel cross-correlation matrix is approximated so that it can be estimated and inverted in real-time.

The convolutive blind signal separation algorithm is based on second order statistics and is entitled 'CoBliSS'. It is implemented in the frequency domain to obtain both a low computational complexity and fast convergence properties. CoBliSS has been tested on synthetically mixed and real recorded speech and music signals using 2 or 3 microphones. Although the separated outputs typically still contain audible crosstalk, the improvement in terms of intelligibility is evident. An extension of CoBliSS is also derived which can simultaneously deal with both signal separation and echo cancellation at a modest increase in computational complexity. CoBliSS is shown to be suited for real-time implementation.

10.2. Open Issues

The performance of both acoustic echo cancellation and blind signal separation in consumer applications significantly depends on implementation aspects. For acoustic echo cancellation a lot of techniques have been used for this purpose including double talk detection and noise reduction by spectral subtraction. Blind signal separation in audio applications is relatively new so that it is mainly in a research and prototyping phase. It is appearing in the first applications at the present time. Similar to the echo canceler, blind signal separation implementation details include detection of speaker presence and noise reduction. Speaker detection for blind signal separation is a difficult problem. The number of total speakers needs to be estimated together with which of these speakers are active at a given time. Noise reduction is also crucial for blind signal separation. In echo cancellation noise is removed merely to obtain a signal with a higher signal to noise ratio. It does not however significantly affect the performance of the echo canceler as the noise is independent of the echo signal. A blind signal separator is however severely hampered by microphone noise. The sensitivity to noise is examined intensively in communications, but there are only few publications that deal with this problem for audio applications, so that this remains an open issue. Some of the publications that take noise into account are Comon, 1996; Moreau and Pesquet, 1997; Moulines et al., 1997; Parra and Spence, 1999. Many other implementation aspects may be important however, a few are mentioned in the next section.

Blind signal separation algorithms that are presented in the literature typically aim at recovering all sources. In the cocktail party problem, humans focus their attention on one particular speaker and neglect all others. When designing a hearing device that uses blind signal separation, this can be obtained by recovering all sources and choosing the desired one. When there are many sources present however, blind signal separation algorithms will not only have a tremendous computational load, they will also perform less accurate or will not separate at all in the limiting case. This is because the number of unknown parameters in the separation system increases rapidly with the number of microphones so that identifiability is easily lost when independence is all the information that can be used. It therefore remains an open issue how to put all effort in recovering one particular source from a mixture of many. Beamforming or combined beamforming and blind signal separation approaches might be more suitable than blind signal separation alone in this context.

10.3. Suggestions for Future Research

Suggestions for future research are given next, based on the work presented in this book and the open issues that are mentioned in the previous section. These suggestions are based on the observation that blind signal separation in audio is a difficult and comprehensive problem so that all available knowledge must be used to obtain a system that is robust to noise, reverberation, moving speakers, and so on. Clearly there is far more information available in practical applications than the mutual independence of the sources alone. Items that are promising for improving blind signal separation in speech applications are

- developing a speaker detector as mentioned in the previous section. This speaker detector can be used to make more reliable cross-correlation estimates. Also, it can automatically pick the separated signal that corresponds to the active speaker. When speakers are not active, finding the signal separation filters can be improved by nulling their corresponding output signal. Also the speaker detector can be used to prevent the separation filters from loosing their focusing on sources that are not active for some time. When no sources are active, such silences can be used for estimating noise properties.

- combining the spectral properties of speech as is used in single channel signal separation with the multichannel blind signal separation that is based on source independence. This can especially lead to significant improvements when the tempo-spatial energies of the sources are far apart, as in a mixture of male and female speech or alternating speakers in general.

- incorporating other speech properties into the blind signal separation algorithm. Techniques and information that is commonly used in the speech recognition area can be used to aid BSS of speech too. These include speech recognition confidence indicators and dictionaries.

- using information that is obtained by other audio components in the system. Although the transfer functions from the speakers to the microphones are not exactly known, the magnitude of these impulse responses as a function of frequency will approximately be equal to that of a transfer function from a loudspeaker to a microphone in the same room. Echo cancelers are typically present in teleconferencing applications so that they can provide the blind signal separator with such information.

- finding a compromise in blind and non-blind systems. Beamforming for example uses the array geometry to separate a desired source from all other sources. When the assumptions on the array and room geometry are not valid however, the performance of a beamformer deteriorates significantly as it relies on this information. Blind signal separation uses no geometric assumptions in general. This makes it more robust but its performance can be further improved for known room and array geometries. A beamforming approach is followed in Li and Sejnowski, 1995. Another method that seems to be promising is presented in Ngo and Bhadkamkar, 1999 where two microphones are spaced only 1 cm apart and a preprocessing is performed that estimates the differences in time of arrival of the sources at the microphones.

Since there will always be a limited amount of test cases as pre-recorded signals, a statistically more reliable view of the performance of an algorithm could be obtained by generating the data according to a speech-like or a music-like distribution, and by doing Monte Carlo experiments over a large number of different realizations of the data. Also, the performance measures can be improved by incorporating perceptual measures, for example.

Developing a real-time prototype is essential for this research however. Working with pre-recorded data sets gives only a limited feel about how the algorithm responds to changes in a real world environment.

Appendix A
Efficient Beamforming

In many implementations of digital delay and sum beamforming, a sample rate much higher than the Nyquist rate is used. This allows for many synchronous beamforming directions. Severe demands are made upon the analog to digital converters however. Several methods have been proposed for reducing the required sample rate. These methods incorporate the delays that are needed for beamforming in the time domain as in Pridham and Mucci, 1978,Schafer and Rabiner, 1973 or in the frequency domain as in Weber and Heisler, 1984. A more efficient method for implementing a time-domain delay and sum beamformer using polyphase decomposition is presented. This method results in significant computational savings when the desired angular resolution is high compared to the number of sensors used and the number of simultaneously formed beams. This work is published in Schobben and Sommen, 1996.

A.1. Introduction

Conventional continuous-time beamformers delay all sensor signals so that propagation delays are canceled and the sensor signals can be combined coherently. In a discrete-time beamformer, these delays are performed digitally. Using discrete time delays only allows for delaying over an integer multiple of the sampling time period. Therefore, the number of synchronous beam-pointing directions is small for low sample rates, resulting in a poor angular resolution. To illustrate this, it is shown in Section A.2 that a linear array sampled at v times the Nyquist rate can only be steered to $1 + 2v$ synchronous angles. A signal arriving from a non-synchronous direction can be received by steering the beam to the most nearby synchronous angle or by rounding the delays needed for beamforming to the delays available. Both methods introduce severe distortion and poor spatial discrimination for small v. In Section A.3 the concept of interpolation beamforming is discussed. This technique uses interpolation, so that the sampling rate is increased artificially. In this way, delays are obtained which are a fractions of the unit delay, see Laakso et al., 1996. In Section A.4 an efficient method is presented for the implementation of the interpolation beamformer using polyphase decomposition. The resulting complexity is discussed in Section A.5 and a numerical example is shown in Section A.6.

A.2. Linear Sensor Array Beamforming

Although the method to be presented can be applied to all array geometries, an linear array example is used. For a linear array the anticipated propagation delay of a flat wave plane from the first sensor to the ith sensor equals

$$\tau_{\max} - \tau_i = \frac{id\sin\theta}{c}, \tag{A.1}$$

where d is the sensor inter-distance, θ is the direction of arrival (DOA) relative to broadside (the direction perpendicular to the line-array), c is the wave plane propagation velocity and $\tau_{\max}$ is the propagation delay of the wave plane between the first and the last sensor. The wave plane incident to the linear array is depicted in Figure A.1. The sensor outputs are sampled at the rate

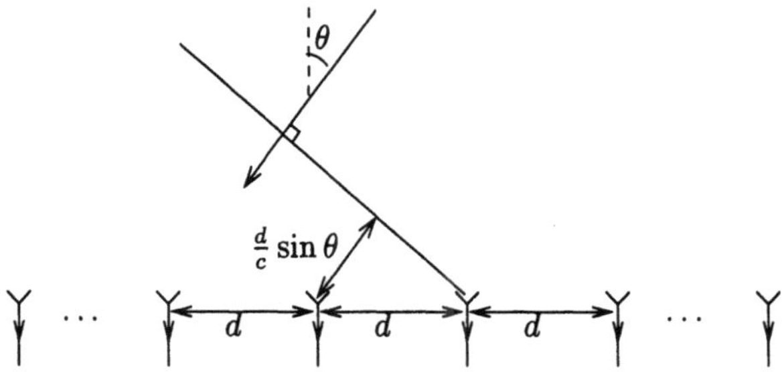

Figure A.1. Wave front incident to a linear array

$f_s = \frac{1}{T_s}$ and consequently the beam can only be steered to the angles which yield a difference in propagation delay of uT_s seconds between neighboring sensors, with u an integer. This delay is canceled using time-delays $uT_s = \frac{d\sin\theta}{c}$. The beam can thus be steered to

$$\theta_u = \sin^{-1}\left(\frac{cuT_s}{d}\right). \tag{A.2}$$

A common choice for d that prevents spatial aliasing is $d = \frac{\lambda_0}{2}$, λ_0 being the minimum wavelength of the signal to be received and f_0 is the corresponding frequency. For this sensor inter-distance and a sample rate equaling $f_s = 2vf_0$, it follows from (A.2) that $|u| \leq v$. There exist $1+2\lfloor v \rfloor$ different u that obey this equation, so the beam can be steered to $1+2\lfloor v \rfloor$ different angles. For example, when the sample rate equals $4f_0$ ($v = 2$), it follows from (A.2) that the beam can only be steered to $0, \pm 30^o$ and $\pm\ 90^o$. A higher angular resolution can be achieved by interpolating the sampled data. When $v = 2$ and more than 5 synchronous directions are desired, the data can be interpolated by a factor K. This is depicted in Figure A.2 for $K = 4$. The solid lines indicate the sampled data and the dashed lines indicate the interpolated data. A

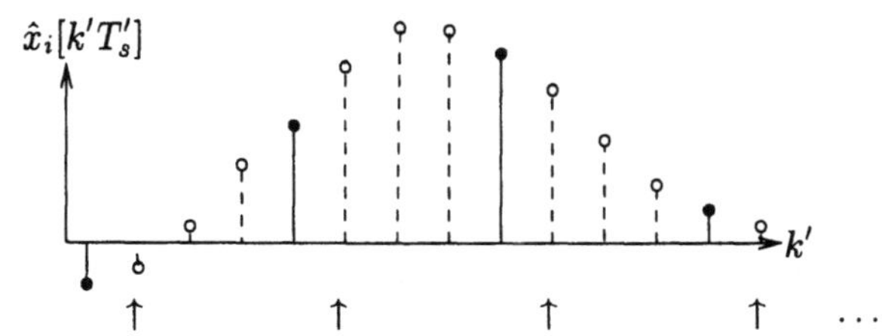

Figure A.2. Data interpolation

delay of $\frac{3T_s}{4}$ for example can now be achieved by selecting the interpolated samples indicated with an ↑ in Figure A.2. Interpolating with a factor $K = 4$ now allows for beamsteering to $0, \pm 7.18^o, \pm 14.5^o, \pm 22.0^o, \pm 30.0^o, \pm 38.7^o, \pm 48.6^o, \pm 60.0^o, \pm 90.0^o$. Clearly, only one of every K interpolated samples is used for beamforming.

A.3. Interpolation Beamforming

The interpolation process for a single beam is depicted in Figure A.3. First the sensor data is sampled at a rate equal to or exceeding the Nyquist sampling rate. The ith sampled sensor output $x_i[kT_s]$ is zero padded to obtain $\tilde{x}_i[k'T'_s]$

$$\tilde{x}_i[k'T'_s] = \begin{cases} x_i\left[\frac{k'}{K}T_s\right], & \text{for } k' = 0, \pm K, \pm 2K, \ldots \\ 0, & \text{otherwise} \end{cases}, \tag{A.3}$$

with $T'_s = \frac{T_s}{K}$. Then the $\tilde{x}_i$ are filtered with the interpolation filter H to obtain the $\hat{x}_i$

$$\hat{x}_i[k'T'_s] = \sum_{l=0}^{N-1} h[lT'_s]\tilde{x}_i[(k'-l)T'_s]. \tag{A.4}$$

This filter is a Finite Impulse Response filter with impulse response $h[k'T'_s]$. The $\hat{x}_i$ are delayed over $p_iT'_s$ seconds, to compensate for the anticipated propagation delays, where the $p_iT'_s$ are equal to the τ_i. Then down-sampling is used to obtain

$$x'_i[kT_s] = \hat{x}_i[(kK - p_i)T'_s]. \tag{A.5}$$

The beamformer output $y[kT_s]$ is obtained by summing the shaded x'_i. Shading means multiplying the x'_i with weights to enhance the angular discrimination. To simplify notation, the shading is not mentioned explicitly in figures and equations. Multiple beams can be formed from the interpolated sensor outputs $\hat{x}_i$ without performing additional multiplications.

In Pridham and Mucci, 1978 it is argued that the scheme of Figure A.3 is equal to the scheme in Figure A.4 for the case that only one beam is formed. This can be seen by interchanging the filter H and the delays $p_iT'_s$ and placing the filter H and the down-sampling in Figure A.3 after the summation. This is allowed when all filters are identical, linear and time invariant. Furthermore, the filter and the down-sampling may be combined to reduce complexity. It will be shown in Section A.5 that the complexity of the technique proposed in the following section is lower than that of the scheme in Figure A.4 for high angular precision beamformers. Note that forming multiple beams is not possible without performing additional multiplications with this scheme.

A.4. Polyphase Decomposition

In this section an efficient implementation of the interpolation beamformer is presented. First, consider the data processing in Figure A.3 for i^{th} sensor only. This is depicted in Figure A.5(a). Here the delay $p_iT'_s$ is interchanged with interpolation filter H. This is allowed since H is linear and time invariant. In Figure A.5(b) filter H is decomposed into the filters $H_0, H_1 \ldots H_{K-1}$ using polyphase decomposition, see Akansu and Haddad, 1992. The impulse responses of the H_j can be calculated from the impulse response of H according to

$$h_j[k'T'_s] = \begin{cases} h[(k'+j)T'_s], & \text{for } k' \in \kappa \\ 0, & \text{otherwise} \end{cases}, \tag{A.6}$$

for $j = 0, 1, \ldots, K-1$ and $\kappa = \{0, K, \ldots, K\lfloor\frac{N-j-1}{K}\rfloor\}$. Down-sampling these filter outputs is equivalent to down-sampling the data and then filtering with H_j', as depicted in Figure A.5(c). The impulse responses of the H_j' are given by

$$h_j'[kT_s] = h_j[kKT_s'], \tag{A.7}$$

for $k = 0, 1, \ldots, \lfloor\frac{N-j-1}{K}\rfloor$. Up-sampling with a factor K, delaying over p_iT_s' and down-sampling with a factor K is equal to delaying over $\frac{p_i}{K}T_s$ if p_i is an integer multiple of K, and equal to zero otherwise. Therefore Figure A.5(c) can be interpreted as choosing the filter H_q' and delaying over rT_s, where

$$\begin{aligned} q &= K\left\lceil\frac{p_i}{K}\right\rceil - p_i \\ r &= \frac{q+p_i}{K} = \left\lceil\frac{p_i}{K}\right\rceil . \end{aligned} \tag{A.8}$$

The resulting scheme is depicted Figure A.5(d). The combination of the delay of rT_s and the sub-filter H_q' represents an approximation of the desired delay p_iT_s'. The sub-filter must be of sufficient length to guarantee a good approximation of the desired delay.

A.5. Comparison of Computational Complexity

As a measure of complexity, the number of multiplications per seconds of the interpolating filter is considered. For simplicity, the filter H is assumed to equal an integer times the interpolation factor K throughout this section. The proposed beamformer is compared with the beamformers of Figure A.3 and A.4.

In the scheme of Figure A.3, J filters of length N are calculated at a rate f_s', with J the number of sensors. Using $f_s' = Kf_s$ and taking advantage of the sparse input data of the interpolation filters, this yields a complexity of NJf_s multiplications per second. This complexity is independent of the number of beams to be calculated, and is therefore efficient when a large number of beams is required. Furthermore, assuming that the filter H is a linear phase filter, the number of multiplications per second can be reduced by approximately a factor 2, yielding a complexity of $\frac{NJf_s}{2}$ multiplications per second.

Filter H in Figure A.4 is calculated for each beam at a rate f_s, since only one of every K samples is needed. The number of multiplications per second equals NJ_Bf_s, with J_B the number of beams to be calculated. Again, assuming that the interpolation filter is a linear phase filter, the resulting number of multiplications per second equals $\frac{NJ_Bf_s}{2}$.

The scheme in Figure A.4 is more efficient than the scheme in Figure A.3 if and only if the number of beams to be calculated is smaller than the number of sensors ($J_B < J$).

In the proposed beamformer, sub-filters H_q' are of length N_s. For each beam, J sub-filters are calculated at a rate f_s. The total number of multiplications per second equals $N_sJJ_Bf_s = \frac{NJJ_Bf_s}{K}$. In general, it is not possible to exploit the linear phase property of the interpolation filter to reduce the complexity further. When $K < 2J_B$ and $J < J_B$ interpolating all sensor data as in Figure A.3 is most efficient. Combining all interpolation filters as in Figure A.4 is most efficient when $K < 2J$ and $J > J_B$. Thus the proposed method outperforms its alternatives when a high angular resolution is required ($K > 2J$ and $K > 2J_B$). This scheme has a gain in computational complexity over its alternatives of $\frac{K}{2J}$ and $\frac{K}{2J_B}$ respectively. The quality of the delays formed depends on the sub-filter length $N_s = \frac{N}{K}$. When both the filter length N and the interpolation factor K are increased with the same factor, the number of beam-pointing directions further increases while the computational complexity does not increase for the proposed method. Consequently, the angles for which beams can be formed can now be chosen with arbitrary precision while maintaining the same amount of multiplications per second. In practice, the filter length N is limited however since $\frac{N}{2}$ tap weights are stored into a finite amount of memory. For the two alternative methods the computational complexity does however increase proportionally with the filter length N.

A.6. Example of Computational Complexity

Next, an example is given to show that the conditions for the proposed method to be more efficient than its alternatives are easily met. Consider sub-filters of length $N_s = 10$, $J = 7$ sensors, $K = 20$ (41 different synchronous angles) and $J_B = 5$ (five beams are formed). The main lobes of the unshaded beam-patterns corresponding to the resulting synchronous beam-pointing directions are depicted in Figure A.6 for θ in between 0° and 90°. For negative angles, the figure is symmetrical. The figure shows that there indeed is a need for a high K to exploit the best possible angular discrimination. However, when K is chosen much larger, the angular discrimination no longer improves, as the successive beams merely overlap. In practice, K will be in between 15 and 40 for a 7 sensor array which is sampled at the Nyquist rate, and the proposed method outperforms the alternatives discussed. The complexity for the proposed beamformer equals $N_s J J_B f_s = 350 f_s$ multiplications per second for this example. The alternate schemes of Figure A.3 and A.4 require $\frac{NJf_s}{2} = 700fs$ and $\frac{NJ_Bf_s}{2} = 500f_s$ multiplications per second respectively. A significant efficiency gain is thus obtained.

A.7. Conclusions

A method using polyphase decomposition is proposed for reduced complexity interpolation delay and sum beamforming. Significant computational savings are reported for beamformers with a high angular discrimination.

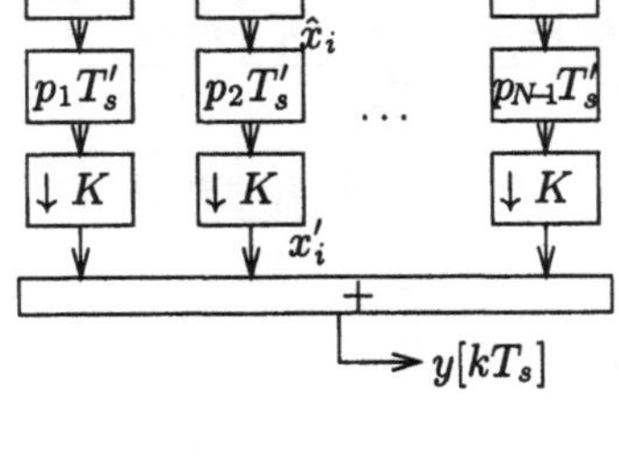

Figure A.3. Interpolation beamformer

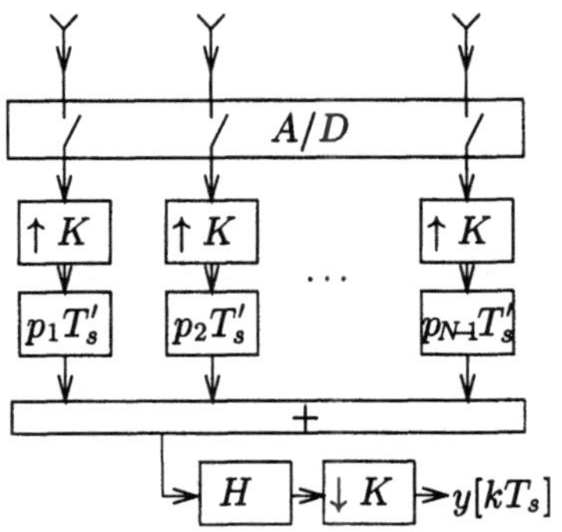

Figure A.4. Beamformer equivalent

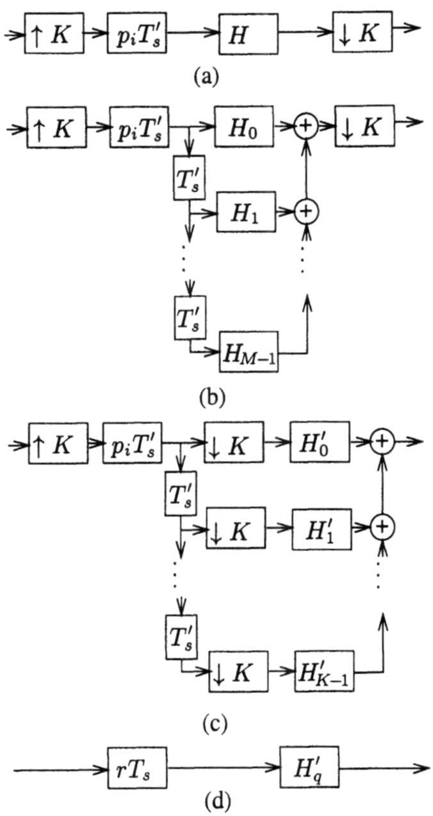

Figure A.5. Interpolation filter (a), polyphase decomposed filter (b), polyphase equivalent (c) and compacted polyphase equivalent (d)

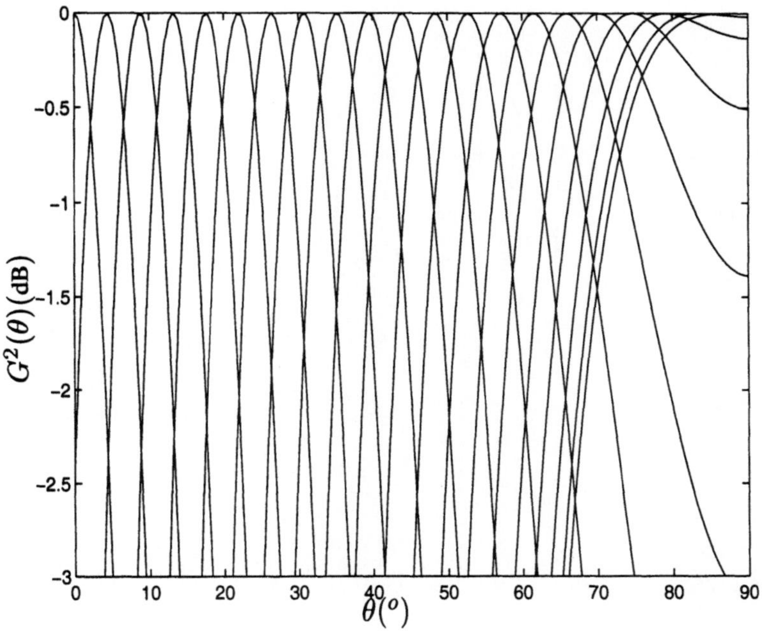

Figure A.6. Beam-pointing patterns for all synchronous directions for a 7 sensor array that is interpolated 20 times

Appendix B
On the Indeterminacies of Convolutive Blind Signal Separation based on Decorrelation

In this appendix conditions are derived for second order statistics to be sufficient in convolutive blind signal separation which are based on mathematical proof and properties of real world signals. A theorem is given for second order statistics to be sufficient for the blind separation of signals that have long autocorrelation sequences compared to the length of the mixing and unmixing filters. For signals that do not have autocorrelations of such lengths it is argued that blind signals separation is still possible given that these signals are non-stationary. Background material and practical implications of this appendix are included in Section 6.2.

B.1. Convolutive BSS Indeterminacies

In this section the uniqueness of the unmixing system is considered taking into account that the unmixing filters are of finite length. The transfer function of the cascaded mixing and unmixing system is also assumed to be of some finite length L and is expressed as

$$y_i[n] = \sum_{l=1}^{J} \sum_{m=0}^{L-1} a_{il}[m] s_l[n-m]\,,$$

with $a_{il}[m]$ the impulse response of the cascaded mixing and unmixing system from source $s_l[n]$ to output $y_i[n]$. The cross-correlations $\rho_{y_i y_j}[\tau] = E\{y_i[n]y_j[n-\tau]\}$ can be grouped into one equation for the correlation lags considered:

$$\begin{pmatrix} \rho_{y_i y_j}[\tau_1] \\ \vdots \\ \rho_{y_i y_j}[\tau_P] \end{pmatrix} = \Xi^{P,J(2L-1)} \begin{pmatrix} \underline{\widehat{a_{i1} a_{j1}}^{2L-1}} \\ \vdots \\ \underline{\widehat{a_{iJ} a_{jJ}}^{2L-1}} \end{pmatrix} \tag{B.1}$$

with the source correlation matrix $\Xi^{P,J(2L-1)}$ given by

$$\left(\begin{array}{ccc|c|ccc} \rho_{s_1}[\tau_1{-}L] & \ldots & \rho_{s_1}[\tau_1{+}L] & & \rho_{s_J}[\tau_1{-}L] & \ldots & \rho_{s_J}[\tau_1{+}L] \\ \vdots & & \vdots & \ldots & \vdots & & \vdots \\ \rho_{s_1}[\tau_P{-}L] & \ldots & \rho_{s_1}[\tau_P{+}L] & & \rho_{s_J}[\tau_P{-}L] & \ldots & \rho_{s_J}[\tau_P{+}L] \end{array}\right)$$

and with

$$\underline{\widehat{a_{il}a_{jl}}}^{2L-1} = \begin{pmatrix} a_{il}[L-1]a_{jl}[0] \\ \sum_{n=L-2}^{L-1} a_{il}[n]a_{jl}[n-L+2] \\ \vdots \\ \sum_{n=1}^{L-1} a_{il}[n]a_{jl}[n-1] \\ \sum_{n=0}^{L-1} a_{il}[n]a_{jl}[n] \\ \sum_{n=1}^{L-2} a_{il}[n]a_{jl}[n+1] \\ \vdots \\ a_{il}[0]a_{jl}[L-1] \end{pmatrix}. \tag{B.2}$$

The $\underline{\widehat{a_{il}a_{jl}}}^{2L-1}$ is a correlation of $\underline{a}_{il}^L$ and $\underline{a}_{jl}^L$, i.e. a convolution of $\mathrm{J}^L\underline{a}_{il}^L$ and $\underline{a}_{jl}^L$, with J^L the $L \times L$ mirror matrix.

This interpretation reveals that $\underline{\widehat{a_{il}a_{jl}}}^{2L-1}$ equals zero iff $\underline{a}_{il}^L$ or $\underline{a}_{jl}^L$ equals zero. Uncorrelated outputs are sufficient to uniquely define the $\underline{\widehat{a_{il}a_{jl}}}^{2L-1}$ when the source correlation matrix $\Xi^{P,J(2L-1)}$ is of rank $J(2L-1)$. In that case its nullspace is empty so that the only solution is $\forall l, i \neq j : \underline{\widehat{a_{il}a_{jl}}}^{2L-1} = \underline{0}^{2L-1}$.

THEOREM B.1 *An unmixing system that produces uncorrelated non-zero outputs at time lags $\tau = \tau_1 \dots \tau_P$ achieves signal separation if the source correlation matrix $\Xi^{P,J(2L-1)}$ is of rank $J(2L-1)$ for this set of time lags.*

Proof:
The proof follows directly from the next two observations:

Observation 1: uncorrelated outputs
The separation system retrieves mutually uncorrelated outputs if the right hand side of (B.1) equals zero for $i \neq j$. This is the case iff $\forall l : \underline{a}_{il}^L = \underline{0}^L$ or $\underline{a}_{jl}^L = \underline{0}^L$. When source l contributes to a certain output i, e.g. $\underline{a}_{il}^L \neq \underline{0}^L$, it cannot contribute to any other output as $\underline{a}_{jl}^L = \underline{0}^L, j \neq i$.

Observation 2: non-zero outputs
To prevent the system from setting the filters in the unmixing system to zero it is required that the autocorrelations of the outputs are non-zero, $\forall i : \exists \tau : \rho_{y_i}[\tau] \neq 0$. This requirement is met when $\forall i : \exists l : \underline{\widehat{a_{il}a_{il}}}^{2L-1} \neq \underline{0}^{2L-1}$ or similarly $\forall i : \exists l : \underline{a}_{il}^L \neq \underline{0}^L$. This means that there is at least one non-zero transfer function $\underline{a}_{il}^L$ to every output i.

According to observation 2 there is a source that contributes to a particular output. According to observation 1 this source cannot contribute to any other outputs. Each source therefore uniquely corresponds to one output □.

B.2. Rank of the Source Correlation Matrix

According to Theorem 1 the source correlation matrix $\Xi^{P,J(2L-1)}$ must be of rank $J(2L-1)$ in order for BSS to succeed using SOS. The number of time lags that are considered P must therefore at least be equal to $J(2L-1)$. Whether the source correlation matrix $\Xi^{P,J(2L-1)}$ is full rank then still depends on

- **Source similarity:** when the autocorrelations of two sources are similar, columns in the source correlation matrix can be eliminated and BSS is not possible in accordance with Theorem 1.
- **Correlation length:** the autocorrelation of at least one source needs to be of sufficient length. When the sources have autocorrelations equal to zero $\rho_{s_i}[\tau] = 0$ for $|\tau| > K$, then the rank of the source correlation matrix is at most $2L - 1 + 2K$.

The well known fact that white noise sources cannot be recovered using SOS is found from investigating the rank of the source correlation matrix. Another result is that sources with short autocorrelation sequences (e.g. moving average processes) can be recovered if one of the sources does have a long autocorrelation sequence. So if the correlation sequences are distinct, one source with a long correlation sequence can render the source correlation matrix full rank. The length of this correlation sequence must be such that $2L - 1 + 2K \geq J(2L - 1)$ which gives

$$K \geq \frac{(J-1)}{2}(2L-1) . \tag{B.3}$$

When there are many sources J, there must be a source with a long correlation sequence in order for SOS-based BSS to work. In audio applications this criterion is not met in general as the correlation length K for signals such as speech is shorter than the length of the cascaded mixing and unmixing system L. Despite of this, successful separation is achieved for several algorithms. This is due to the non-stationarity of the data and implementation details of such algorithms as discussed in the next section.

B.3. Real World Data Considerations

Source signals that have short autocorrelation sequences can be whitened when long unmixing filters are used. BSS is therefore not possible when using SOS, which is in correspondence with (B.3). However, typically constraints are incorporated into BSS algorithms to prevent them from whitening the data. For audio signals this is usually done so that the separated signals sound more natural. It has as a side effect however that the algorithms are forced to separate as they are prevented from minimizing their cost-function by whitening the data.

Another important aspect is that most analysis - including the one in the previous section - assume stationarity of the source signals. Real world applications however typically involve the separation of signals that are not stationary. Consider for example speech signals which are recorded in a real room. Speech signals are stationary only over approximately 10-30 milliseconds. Thus, speech can be modeled by time varying innovation filters excited by white noise. These innovation filters change therefore every 10-30 ms. Typically, the convergence time of the SOS algorithm is slower than 10-30 ms so that the BSS algorithm fails to track the inverse of the innovation filters and only true separation results in uncorrelated recovered signals at all times. This is in correspondence with (6.5) in which for each choice of stationary source signals there exist a set of unmixing filters that renders the cross-correlations among the outputs equal to zero. This set is different for other stationary source signals but it always contains the desired solution that gives true signal separation. This true solution can therefore be found from the intersection of the solution sets. This is obtained automatically in an adaptive system when the stationarity of the source signals is over a shorter time than the convergence time of the adaptive system.

Appendix C
Computational Complexity of CoBliSS

In this Appendix the computational complexity of all parts of the CoBliSS algorithm is derived. The properties of the matrices involved are exploited to obtain an efficient implementation. It is assumed that a complex multiplication is performed using 4 real multiplications (muls) and 2 real additions (adds). As a measure of complexity, the number of floating point operations is used which is assumed to be equal to the sum of real muls and adds. In the Fourier domain, there are $M/2-1$ frequency bins with complex data and 2 frequency bins with real data. In the sequel $M/2$ complex bins are considered instead for the sake of simplicity. This is justified as in typical cases M is large, e.g. $M = 2048$, and J is small, e.g. $J = 2$.

C.1. Fast Fourier Transforms

In the CoBliSS algorithm, windowed Fast Fourier Transforms (FFTs) can be used. The computational complexity of the FFTs that transform the data to the frequency domain is $J\frac{5}{2}M\log_2(B)$ flops as derived in Chapter 4. The same complexity is needed for transforming the filtered data back to the time domain, as only B output samples are relevant. Also, $2J^2$ (I)FFTs are needed to constrain the filters to be of finite length. This requires $2J^2\frac{5}{2}M\log_2(N)$ flops. For CoBliSS, $B = N$ is a common choice so that the (I)FFTs require a total of $(J^2+J)5M\log_2(N)$ flops.

C.2. Correlation Update

In the calculation of the correlation update, two aspects are exploited. One is that the correlation update

$$\underline{P}_{ac}^{M} := \alpha\underline{P}_{ac}^{M} + (1-\alpha)\left((\underline{X}_{a}^{M})^{*}\otimes\underline{X}_{c}^{M}\right)$$

is symmetrical, i.e. $\underline{P}_{ac}^{M} = (\underline{P}_{ca}^{M})^{*}$, as can be seen from its definition. The other is that it is real for $a = c$. The complexity required to element-wise multiply the $\frac{J(J-1)}{2}\frac{M}{2}$ complex elements $(a \neq c)$ is $\frac{J(J-1)}{2}\frac{M}{2}$ complex muls. The complexity required to compute the $J\frac{M}{2}$ real elements $(a = c)$ is $2J\frac{M}{2}$ real muls and $J\frac{M}{2}$ real adds.

Multiplying with $(1-\alpha)$ takes $J\frac{M}{2}$ real muls for the real elements $(a = c)$ and $\frac{J(J-1)}{2}2\frac{M}{2}$ real muls for the complex elements $(a \neq c)$. The same complexity is required to calculate the product $\alpha\underline{P}_{ac}^{M}$. Also, the results must be added, which takes $J\frac{M}{2} + \frac{J(J-1)}{2}2\frac{M}{2}$ real adds.

In total, $2J^2M$ real muls and J^2M real adds are required for the correlation update.

C.3. Weight Update

A computational expensive part of the algorithm is the matrix square root that is required for the weight update. However, once the estimations of the cross-correlation matrices $\underline{P}_{ac}^{M}$ converge, the weight update can be approximated as derived in Subsection 7.5 by

$$W_p'^J = W_p^J(\mathrm{I}^J - \tfrac{1}{2}(\Delta P_p^J)^*(W_p^J)^H W_p^J) \tag{C.1}$$

with ΔP_p is the updated P_p minus the previous estimate of P_p.

In the fast weight update the symmetrical matrix $(W_p^J)^H W_p^J$ is calculated which is multiplied with the symmetrical matrix $(\Delta P_p^J)^*$. The multiplication with $\frac{1}{2}$ can readily be combined with $(1-\alpha)$ in the correlation update. The result is subtracted from the identity matrix and then multiplied with W_p^J.

The product $(W_p^J)^H W_p^J$ can be written as

$$((W_p^J)^H W_p^J)_{k,l} = \sum_{m=1}^{J} (W_p^J)^*_{m,k}(W_p^J)_{m,l}.$$

For the diagonal elements ($k = l$), $2J^2$ real muls and $2J^2 - J$ real adds are required. For the off-diagonal elements ($k \neq l$), $\frac{J(J^2-J)}{2}$ complex muls and $(J-1)\frac{J^2-J}{2}$ complex adds are required. In total, $2J^3$ real muls and $2J^3 - J^2$ real adds are needed to calculated $(W_p^J)^H W_p^J$.

The matrix subtraction that yields ΔP_p takes J real adds for the diagonal elements and $J\frac{(J-1)}{2}$ complex adds for the off-diagonal elements; J^2 real adds in total. The multiplication of the symmetrical matrices $(\Delta P_p^J)^*$ and $(W_p^J)^H W_p^J$ takes $4J^3 - 6J^2 + 3J$ real muls and $4J^3 - 7J^2 + 2J$ real adds[1]. To subtract the result from the identity matrix takes J real adds. In total, the weight update requires $6J^3 - 6J^2 + 3J$ real muls and $6J^3 - 7J^2 + 2J$ real adds.

C.4. Filtering

The filtering is performed efficiently in the frequency domain using the overlap-save technique. Besides the (I)FFTs, the filtering requires element-wise vector products. This takes $\frac{M}{2}J^2$ complex muls and $\frac{M}{2}(J-1)J$ complex adds.

C.5. Normalization

The W_p are normalized to prevent them from collapsing due to extreme high or low energy of the input signals for frequency bin p. To calculate the l_2 norm of W_p, $2J^2$ real muls are needed to obtain the magnitude of the elements of W_p. Also, J^2 real adds are needed to add these magnitudes and one square root is required to calculate the norm. All elements of W_p must be divided by the norm, which takes $2J^2$ real muls. In total, the normalization takes $4J^2$ real muls and $J^2 - 1$ real adds and 1 square root. The weight update and the normalization must be done for all $M/2$ frequency bins.

Notes

1 The product of two complex symmetrical matrices can be written as

$$c_{kl} = \sum_{m=1}^{J} a_{km} b_{ml} \text{ with } a_{km} = a_{mk}^{*} \text{ and } b_{ml} = b_{lm}^{*}$$

In the calculation of each of the $J^2 - J$ off-diagonal elements there are 2 real numbers involved (a_{kk} and b_{kk} are real $\forall k$) so that only $J - 2$ complex muls and 4 real muls are required. Also, $J - 1$ complex additions are needed for the summation. In the calculation of each of the J diagonal elements one real mul and $J - 1$ complex muls are needed. All complex muls are encountered twice in the calculation of the diagonal elements, so that only half of these need to be computed. Also, J real adds and $(J-2)J$ complex adds are needed to sum the results. In total, $4J^3 - 6J^2 + 3J$ real muls and $4J^3 - 7J^2 + 2J$ real adds are needed. If the matrix symmetries would not have been exploited, a total of $4J^3$ real muls and $4J^3 - 2J^2$ would have been required. Hence, a significant improvement is achieved for small J.

Appendix D
Efficient computation of windowed FFTs

This appendix summarizes a computationally efficient implementation of the windowed (I)FFT and a method to compute the Fourier transform of overlapping data using a windowed FFT as is shown in Schobben et al., 1997 and Egelmeers, 1995.

D.1. Efficient Computation of Windowed (I)FFTs

Windowed FFTs transform a block of Q input samples which is augmented with zeros. Windowed IFFTs are inverse transforms of which only a block of Q outputs is needed. Both the FFT length M and the window length Q are assumed to be powers of 2. The FFTs transform real data and the IFFTs generate real data. The number of real floating point operations (flops), i.e. the sum of the real multiplications and additions, will be considered as a measure of computational complexity. It is assumed that 4 real multiplications and 2 real additions are used for 1 complex multiplication. The number of flops needed to calculate a length M FFT and IFFT with window length Q is denoted as $\Psi_{\mathrm{FFT}}\{M, Q\}$ and $\Psi_{\mathrm{IFFT}}\{M, Q\}$ respectively. The windowed FFT of real input data is defined as

$$\underline{X}^M = \mathcal{F}^M \begin{pmatrix} \underline{0}^{M-Q} \\ \underline{x}^Q \end{pmatrix}, \tag{D.1}$$

with $\underline{x}^Q$ defined by (4.2) with $B = Q$. Although the window can also be on the other side as in (4.5), modifications will be made to the update in Section 4.5 so that all (I)FFTs are in correspondence with (D.1). Time indices are suppressed to simplify notation. A radix-2 length M Decimation In Time (DIT) FFT decomposition is defined for $0 \leq l < M$ as Oppenheim and Shafer, 1975

$$\begin{aligned}
(\underline{X}^M)_l &= \sum_{k=0}^{M} \begin{pmatrix} \underline{0}^{M-Q} \\ \underline{x}^Q \end{pmatrix}_k e^{-\jmath 2\pi \frac{kl}{M}}, \\
&= \sum_{k=0}^{\frac{M}{2}-1} \begin{pmatrix} \underline{0}^{M-Q} \\ \underline{x}^Q \end{pmatrix}_{2k} e^{-\jmath 2\pi \frac{kl}{M/2}}
\end{aligned}$$

$$+\ e^{-\jmath 2\pi \frac{l}{M}} \sum_{k=0}^{\frac{M}{2}-1} \begin{pmatrix} \underline{0}^{M-Q} \\ \underline{x}^{Q} \end{pmatrix}_{2k+1} e^{-\jmath 2\pi \frac{kl}{M/2}}\,, \tag{D.2}$$

where the subscript denotes the element that is taken from the corresponding vector. The number of flops needed to combine these two FFTs is denoted as $\Psi_C\{M\}$. The total number of flops equals for $M > 4$ and $Q > 2$

$$\Psi_{\text{FFT}}\{M,Q\} = 2\Psi_{\text{FFT}}\{\tfrac{M}{2},\tfrac{Q}{2}\} + \Psi_C\{M\}\,. \tag{D.3}$$

This means that a windowed FFT is split in two length $\frac{M}{2}$ FFTs with window length $\frac{Q}{2}$. The combining is achieved by multiplying the result of one FFT element-wise with elements of the Fourier matrix $\mathcal{F}^M$ and adding this result to the result of the other FFT. This requires[1]

$$\Psi_C\{M\} = \tfrac{5}{2}M - 8 \tag{D.4}$$

flops. The FFTs are split until FFTs of length $\frac{2M}{Q}$ remain with windows of length 2. The number of flops needed to calculate such a FFT is[2]

$$\Psi_{\text{FFT}}\{\tfrac{2M}{Q},2\} = \tfrac{3M}{Q} - 2 \tag{D.5}$$

for $Q < M$. The overall complexity can now be calculated by inserting (D.5) and (D.4) in (D.3) which yields for $Q < M$

$$\begin{aligned}\Psi_{\text{FFT}}\{M,Q\} &= \tfrac{Q}{2}\Psi_{\text{FFT}}\{\tfrac{2M}{Q},2\} + \sum_{b=1}^{\log_2(\frac{Q}{2})} 2^{b-1}\Psi_{\text{C}}\{\tfrac{M}{2^{b-1}}\} \\ &= \tfrac{5}{2}M\log_2(Q) - M - 5Q + 8\,. \end{aligned} \tag{D.6}$$

When calculating IFFTs, the complexities for combining two length $\frac{M}{2}$ IFFTs and performing a length $\frac{2M}{Q}$ IFFT of which only 2 outputs are needed are

$$\Psi_{IC}\{M\} = \tfrac{5}{2}M - 6 \ \text{ and } \ \Psi_{\text{IFFT}}\{\tfrac{2M}{Q},2\} = \frac{8M}{Q} - 2\,.$$

Following a similar procedure as for the FFT, the overall complexity of the IFFT is

$$\Psi_{\text{IFFT}}\{M,Q\} = \tfrac{5}{2}M\log_2(Q) + \tfrac{3}{2}M - 5Q + 6 \tag{D.7}$$

for $Q < \frac{M}{2}$. Note that the FFT can be implemented with even lower complexity for $Q = M$ than that indicated by (D.6) by starting with lengths 4 FFT which requires $\Psi_{\text{FFT}}\{4,4\} = 6$ flops. Also, a more efficient implementation of the IFFT is achieved for $Q = M$ and $Q = \frac{M}{2}$ by starting with length 4 IFFTs which requires $\Psi_{\text{IFFT}}\{4,4\} = 8$ and $\Psi_{\text{IFFT}}\{4,2\} = 6$ flops respectively.

D.2. Efficient Recursive FFTs of Overlapping Data

Four (I)FFTs in the BFDAF can be calculated using the windowed procedure described above. The FFT which transforms the overlapping input data does not contain a window and will therefore be the largest computational burden. In this section a method is presented which performs the FFT of the overlapping input data using a windowed FFT with window length B. The overlapping input data at time $(k+1)B$ can be written as

$$\underline{x}^M[(k+1)B] = D_B^M \underline{x}^M[kB] + \underline{y}_B^M[kB],$$

with $\underline{y}_B^M[kB]$ the windowed data vector

$$\underline{y}_B^M[kB] = \begin{pmatrix} \underline{0}^{M-B} \\ \underline{x}^B[(k+1)B] - \underline{x}^B[kB-M+B] \end{pmatrix}$$

and D_B^M a circular shift defined as

$$\mathrm{D}_B^M = \begin{pmatrix} 0^{M-B,B} & \mathrm{I}^{M-B} \\ \mathrm{I}^B & 0^{B,M-B} \end{pmatrix}.$$

Now the transform of the overlapping input data at time $(k+1)B$ can be written as

$$\begin{aligned} \underline{X}^M[(k+1)B] &= \mathcal{F}^M(D_B^M \underline{x}^M[kB] + \underline{y}_B^M[kB]) \\ &= \mathcal{F}^M D_B^M (\mathcal{F}^M)^{-1} \underline{X}^M[kB] + \mathcal{F}^M \underline{y}_B^M[kB]. \end{aligned} \qquad \text{(D.8)}$$

Thus, the transform of the overlapping input data at time $(k+1)B$ can be calculated from the sum of previously transformed data and the result of a windowed FFT with window length B. Also, a circular shift $\mathcal{F}^M D_B^M (\mathcal{F}^M)^{-1}$ must be performed. The resulting computational complexity therefore equals $\Psi_{\mathrm{FFT}}\{M,B\} + \Psi_{\mathrm{ROT}}\{M,B\}$, with $\Psi_{\mathrm{ROT}}\{M,B\}$ the number of flops needed for the circular shift. Using the circular structure of D_B^M yields $\mathrm{D}_B^M = (\mathrm{D}_1^M)^B$ Davis, 1979. This circular shift can be computed efficiently in the frequency domain as

$$\mathcal{F}^M \mathrm{D}_1^M (\mathcal{F}^M)^{-1} = \mathrm{diag}\{\underline{v}^M\}$$

with

$$\underline{v}^M = \mathcal{F}^M \begin{pmatrix} \underline{0}^{M-1} \\ 1 \end{pmatrix}.$$

As $\mathcal{F}^M \mathrm{D}_B^M (\mathcal{F}^M)^{-1} = (\mathcal{F}^M \mathrm{D}_1^M (\mathcal{F}^M)^{-1})^B$, the k'th element of the diagonal of $\mathcal{F}^M \mathrm{D}_B^M (\mathcal{F}^M)^{-1}$ equals $((\underline{v}^M)_k)^B$. The diagonal matrix $\mathcal{F}^M \mathrm{D}_B^M (\mathcal{F}^M)^{-1}$ can now be written as

$$\mathcal{F}^M \mathrm{D}_B^M (\mathcal{F}^M)^{-1} = \mathrm{diag}\{\begin{pmatrix} \underline{v}^{\frac{M}{B}} \\ \vdots \\ \underline{v}^{\frac{M}{B}} \end{pmatrix}\}.$$

Thus, the circular shift reduces to an element-wise multiplication in (D.8). Using that the circular shift contains B length M/B Fourier vectors and that both the left and right operand of the element-wise multiplication are Fourier transforms of real valued vectors the total circular shift requires $\Psi_{\mathrm{ROT}}\{M,B\}$ real operations

$$\Psi_{\mathrm{ROT}}\{M,B\} = \begin{cases} 0 & \text{for } \frac{M}{B} < 8 \\ 3M - 16B & \text{for } \frac{M}{B} \geq 8 \end{cases}.$$

Notes

1 $\frac{M}{4} - 1$ complex multiplications are needed for multiplication with $e^{-j2\pi \frac{kl}{M/2}}$ in (D.2). $4(\frac{M}{4} - 1)$ real additions are needed to add the resulting real and imaginary terms to the first part of (D.2). Another 2 real additions are needed for $l = 0$ and $l = \frac{M}{2}$.

2 The FFT is written as

$$X_l = x_{\frac{2M}{Q}-2} e^{j2\pi \frac{2l}{2M/Q}} + x_{\frac{2M}{Q}-1} e^{j2\pi \frac{l}{2M/Q}}.$$

The number of real multiplications in each of both terms that are not equal to 1 or j is $\frac{M}{2Q} - 1$. Another $2(\frac{M}{Q} - 1)$ real adds are needed to add the contributions of both terms. Also, 2 real additions are needed for $l = 0$ and $l = \frac{M}{2}$.

Appendix E
Contents of the Compact Disc

The compact disc is a mixed mode CD-rom. The sound tracks listed below correspond to the experiments and can be played using an audio CD player. They have been resampled at 44.1kHz for this purpose. In addition to the audio tracks, the compact disc contains test data, Matlab code and papers by the author. These files can be accessed using a CD-rom player. Details can be found in the README.txt file.

Efficient Multichannel RLS

Two loudspeakers reproduce two highly correlated signals which are picked up by a microphone. The MC-BFDAF algorithm is used to estimate the contributions from the loudspeaker signals to the microphone. This estimate is subtracted from the microphone signal to obtain the residual (error) signal. To show the improvement due to the joint normalization in the MC-BFDAF, the same experiment is repeated with two separate single-channel BFDAF echo cancelers. Details about this experiment can be found in Section 5.7.

1. Loudspeaker 1 (x_1)
2. Loudspeaker 2 (x_2)
3. Microphone (x)
4. Residual (e)
5. Residual when two separate single channel echo cancelers are used

CoBliSS

Synthetical mixture

The source signals are music signals in this experiment. They are mixed synthetically using measured room impulse responses to obtain the microphone signals. The CoBliSS algorithm processes these microphone signals to obtain the separated output signals. Details about this experiment can be found in Section 7.9.

6. Source 1 (s_1)
7. Source 2 (s_2)
8. Microphone 1 (x_1)
9. Microphone 2 (x_2)

10. Separated output 1 (y_1)
11. Separated output 2 (y_2)

Two male speakers speaking simultaneously

The CoBliSS algorithm recovers two male speech signals from a stereo microphone recording. The source signals are therefore unknown. Details about this experiment can be found in Section 7.9.

12. Microphone 1 (x_1)
13. Microphone 2 (x_2)
14. Separated output 1 (y_1)
15. Separated output 2 (y_2)

Two male speakers speaking simultaneously together with a loudspeaker that reproduces the French radio news

The previous experiment is repeated with an additional loudspeaker present in the room. The loudspeaker reproduces the French radio news and this signal is fed to the CoBliSS algorithm as a third input signal. CoBliSS does not exploit the fact that this signal readily is a source signal. Details can be found in Section 7.9.

16. Microphone 1 (x_1)
17. Microphone 2 (x_2)
18. Radio news (x_3)
19. Separated output 1 (y_1)
20. Separated output 2 (y_2)
21. Separated output 3 (y_3)

ECoBliSS

The data of the previous experiment is processed using the Extended CoBliSS algorithm. The loudspeaker signal readily is a source signal and only the speech signals are recovered. Details about this experiment can be found in Section 8.4.

22. Separated output 1 (y_1)
23. Separated output 2 (y_2)

Appendix F
Glossary

The mathematical notation and symbols are described in detail in Section 1.7. The abbreviations that are used throughout the book are listed below

#	Number of
A/D	Analogue to Digital converter
AEC	Acoustic Echo Canceler
BFDAF	Block Frequency Domain Adaptive Filter
BSS	Blind Signal Separation
CoBliSS	A new Convolutive Blind Signal Separation algorithm based on SOS
D/A	Digital to Analogue converter
DFT	Discrete Fourier Transform
DSP	Digital Signal Processor
ECoBliSS	Extended CoBliSS; a generalization for simultaneous BSS and AEC
FDAF	Frequency Domain Adaptive Filter
FIR	Finite Impulse Response
FFT	Fast Fourier Transform
HOS	Higher Order Statistics
ICA	Independent Component Analysis
iff	If and only if
IDFT	Inverse Discrete Fourier Transform
IFFT	Inverse Fast Fourier Transform
flop	Floating Point Operation
LMS	Least Mean Square
mic	Microphone
MC	Multichannel
MIPS	Million Instructions Per Second
MISO	Multiple Input Single Output
NLMS	Normalized LMS
PDF	Probability Density Function
RLS	Recursive Least Squares
SOS	Second Order Statistics

References

Affes, S. and Grenier, Y. (1997). A signal subspace tracking algorithm for microphone array processing of speech. *IEEE Trans. on Speech and Audio*, 5(5):425–437.

Akansu, A. and Haddad, R. (1992). *Multiresolution signal decomposition*. London : Academic Press.

Alberge, F., Duhamel, P., and Grenier, Y. (1998). A combined fdaf/wsaf algorithm for stereophonic acoustic echo cancellation. In *Int. Conf. on Acoustics, Speech and Signal Processing (ICASSP)*, pages 1413–1416.

Ali, M. (1998). Stereophonic acoustic echo cancellation system using time-varying all-pass filtering for signal decorrelation. In *Int. Conf. on Acoustics, Speech and Signal Processing (ICASSP)*, pages 3689–3692.

Amari, S., Cichocki, A., and Yang, H. H. (1996). A new learning algorithm for blind signal separation. In *Advances in Neural Information Processing Systems 8*, pages 757–763. MIT Press.

Applebaum, S. (1976). Adaptive arrays. *IEEE Trans. on Antennas and Propagation*, AP-24(5):585–598.

Bell, A. and Sejnowski, T. (1995a). An information maximisation approach to blind separation and blind deconvolution. *Neural Computation*, 7(6):1129–1159. MIT Press, Cambridge MA.

Bell, A. and Sejnowski, T. (1995b). An information-maximisation approach to blind separation and blind deconvolution. *Neural Computation*, 7(6):1129–1159.

Belt, H., den Brinker, A., and Benders, F. (1995). Adaptive line enhancement using a second-order iir filter. In *Proc. ICASSP*, pages 1444–1447, Detroit (MI), U.S.A.

Benesty, J., Amand, F., Gilloire, A., and Grenier, Y. (1995). Adaptive filtering algorithms for stereophonic acoustic echo cancellation. In *Int. Conf. on Acoustics, Speech and Signal Processing (ICASSP)*, pages 3099–3102, Detroit.

Benesty, J., Morgan, D., Hall, J., and Sondhi, M. (1998). Stereophonic acoustic echo cancellation using nonlinear transformations and comb filtering. In *Int. Conf. on Acoustics, Speech and Signal Processing (ICASSP)*, pages 3637–3676.

Boll, S. (1979). Suppression of acoustic noise in speech using spectral subtraction. *IEEE Trans. on Acoustics, Speech and Signal Processing*, 27(2):113–120.

Cadzow, J. (1996). Blind deconvolution via cumulant extrema. *IEEE Signal Processing magazine*, 13(3):24–42.

Cardoso, J.-F. (1989). Source separation using higher order moments. In *Proc. ICASSP*, pages 2109–2112, Glasgow, Scotland.

Cardoso, J.-F. and Laheld, B. (1996). Equivariant adaptive source separation. *IEEE Trans. on Signal Processing*, 44(12):3017–3030.

Chan, D. (1997). *Blind signal separation*. Cambridge: Thesis University of Cambridge. Ph. D. Thesis.

Cherry, E. (1953). Some experiments on the recognition of speech with one and with two ears. *Journal of the Acoustical Society of America*, 25:975–979.

Clifford, M. (1986). *Microphones*. Blue Ridge Summit, 3rd ed. edition.

Comon, P. (1994). Independent component analysis – a new concept? *Signal Processing*, 36(3):287–314.

Comon, P. (1996). Contrasts for multichannel blind deconvolution. *IEEE Signal Processing Letters*, 3(7):209–211.

Compton, R. (1988). *Adaptive antennas*. Prentice Hall, Englewood Cliffs, New Jersey.

Cremer, L. and Muller, H. (1982). *Principle and applications of room acoustics*. London: Applied Science Publishers.

Davis, P. (1979). *Circulant matrices*. New York (USA): Wiley.

Delfosse, N. and Loubaton, P. (1996). Adaptive blind separation of convolutive mixtures. In *Int. Conf. on Acoustics, Speech and Signal Processing (ICASSP)*, pages 2940–2943.

Egelmeers, G. (1995). *Real time realization of large adaptive filters*. Eindhoven : Eindhoven Technical University of Technology, The Netherlands. Ph.D. Thesis.

Ferrara, E. (1980). Fast implementation of lms adaptive filters. *IEEE Trans. on Acoustics, Speech, and Signal Processing*, ASSP-28(4):474–475.

Frost, O. (1972). An algorithm for linearly constrained adaptive array processing. *Proc. of the IEEE*, 60(8):926–935.

Garas, J. (1996). *Personal sound using active sound control*. Eindhoven University of Technology. Report Stan Ackermans Institute ; 96014.

Gardner, W. (1995). Efficient convolution without input/output delay. *Journal of the Audio Engineering Society*, 43(3):127–136.

Gillore, A. and Turbin, V. (1998). Using auditory properties to improve the behaviour of stereophonic acoustic echo cancelers. In *Int. Conf. on Acoustics, Speech and Signal Processing (ICASSP)*, pages 3681–3684.

Girolami, M. (1998). Noise reduction and speech enhancement via temporal anti-Hebbian learning. In *Proc. ICASSP*, Seattle, WA, USA.

Griffiths, L. (1969). A simple adaptive algorithm for real-time processing in antenna arrays. *Proc. of the IEEE*, 57(10):1696–1704.

Griffiths, L. and Jim, C. (1982). An alternative approach to linearly constrained adaptive beamforming. *IEEE Trans. on Antennas and Propagation*, AP-30:27–34.

Gu, Y. (1992). *Linear and nonlinear adaptive filtering and their application to speech intelligibility enhancement*. Eindhoven (Netherlands): Eindhoven Technical University of Technology. Ph. D. Thesis.

Haykin, S. (1994). *Blind deconvolution*. Englewood Cliffs : PTR Prentice Hall.

Haykin, S. (1996). *Adaptive filter theory*. Englewood Cliffs : Prentice Hall, 3rd edition.

Haykin, S. (2000a). *Unsupervised adaptive filtering - blind deconvolution*, volume II. Wiley & Sons.

Haykin, S. (2000b). *Unsupervised adaptive filtering - blind source separation*, volume I. Wiley & Sons.

Hirano, A., Nakayama, K., and Watanabe, K. (1999). Convergence analysis of stereophonic echo canceler with pre-processing: Relation between pre-processing and convergence. In *Int. Conf. on Acoustics, Speech and Signal Processing (ICASSP)*.

Ikeda, S. and Murata, N. (1998). An approach to blind source separation of speech signals. *Proc. of ICANN'98*.

Ikeda, S. and Murata, N. (1999). A method of ica in time-frequency domain. *Proc. Int. Workshop Independent Component Analysis and Blind Signal Separation,* pages 365–370.

Jelonnek, B., Boss, D., and Kammeyer, K.-D. (1997). Generalized eigenvector algorithm for blind equalization. *Signal Processing,* 61(3):237–264.

Joho, M. (2000). *A Systematic Approach to Adaptive Algorithms for Multichannel System Identification, Inverse Modeling an Blind Identification.* Hartung-Gorre Verlag, Konstanz. Ph.D. Thesis.

Joho, M. and Moschytz, G. (1997). Adaptive beamforming with partioned frequency-domain filters. In *IEEE Workshop on Applications of Signal Proc. to Audio and Acoustics.*

Joncour, Y. and Sugiyama, A. (1998). A stereo echo canceler with pre-processing for correct echo path identification. In *Int. Conf. on Acoustics, Speech and Signal Processing (ICASSP),* pages 3677–3680.

Jutten, C. and Cardoso, J.-F. (1995). Separation of sources: Really blind? In *Proc. of the Int. Symposium on Nonlinear Theory and Applications (NOLTA),* pages 79–84, Las Vegas, Nevada, USA.

Laakso, T., Valimaki, V., Karjalainen, M., and Laine, U. (1996). Splitting the unit delay. *IEEE Signal Processing Magazine,* 13(1):30–60.

Lambert, R. (1996). *Multichannel blind deconvolution: FIR matrix algebra and separation of multipath mixtures.* University of Southern California. Ph.D. Thesis.

Lambert, R. and Bell, A. (1997). Blind separation of multiple speakers in a multipath environment. In *Int. Conf. on Acoustics, Speech and Signal Processing (ICASSP).*

Lee, T.-W. (1998). *Independent Component Analysis.* Kluwer.

Lee, T.-W., Bell, A., and Lambert, R. (1997a). Blind separation of delayed and convolved sources. *Advances in Neural Inf. Proc. Systems 9,* pages 758–764.

Lee, T.-W., Bell, A., and Orglmeister, R. (1997b). Blind source separation of real world signals. In *Proc. ICNN,* Houston, TX.

Li, S. and Sejnowski, T. J. (1995). Adaptive separation of mixed broad-band sound sources with delays by a beamforming Hérault-Jutten network. *IEEE Journal of Oceanic Engineering,* 20(1):73–79.

Mansour, D. and jr., A. G. (1982). Unconstrained frequency domain adaptive filter. *IEEE Trans. on Acoustics, Speech and Signal Processing,* 30(5):726–734.

Miyoshi, M. and Kaneda, Y. (1988). Inverse filtering of room acoustics. *IEEE Trans. on Acoustics, Speech and Signal Processing,* 36(2):145–152.

Molgedey, L. and Schuster, H. (1994). Separation of independent signals using time-delayed correlations. *Physical Review Letters,* 72(23):3634–3637.

Moreau, E. and Pesquet, J.-C. (1997). Generalized contrasts for multichannel blind deconvolution of linear systems. *IEEE Signal Processing Letters,* 4(6):182–183.

Moulines, E., Cardoso, J.-F., and Gassiat, E. (1997). Maximum likelihood for blind separation and deconvolution of noisy signals using mixture models. In *Proc. ICASSP,* pages 3617–3620, Munich, Germany.

Nelson, P. and Elliott, S. (1992). *Active control of sound.* Academic Press.

Nelson, P., na Bustamante, F. O., and Hamada, H. (1995). Inverse filter design and equalization zones in multichannel sound reproduction. *IEEE Trans. on Speech and Audio Processing,* 3(3):185–192.

Ngo, J. T. and Bhadkamkar, N. A. (1999). Adaptive blind separation of audio sources by a physically compact device using second-order statistics. In *Proc. ICA and BSS,* Aussois, France.

Oppenheim, A. and Shafer, R. (1975). *Digital signal processing.* Englewood Cliffs, New York (USA): Prentice-Hall.

Owens, F. (1993). *Signal processing of speech.* Macmillan.

Parra, L. and Spence, C. (1998). Convolutive blind source separation based on multiple decorrelation. In *Proc. of NNSP98*, Cambridge, UK.

Parra, L. and Spence, C. (1999). Convolutive blind source separation based on multiple decorrelation. *IEEE Trans. on Speech and Audio Processing*. (submitted).

Pillai, S. (1989). *Array signal processing*. Berlin : Springer.

Plomp, R. (1978). Auditory handicap of hearing impairment and the limited benefit of hearing aids. *Journal of the Acoustical Society of America*, 63(2):533–549.

Pope, K. and Bogner, R. (1996a). Blind signal separation: I - linear, instantaneous combinations. *Digital Signal Processing*, 6(2):5–16.

Pope, K. and Bogner, R. (1996b). Blind signal separation: II - linear, convolutive combinations. *Digital Signal Processing*, 6(3):17–28.

Pridham, R. and Mucci, R. (1978). A novel approach to digital beamforming. *Journal of the Acoustical Society of America*, 63:425–434.

Sahlin, H. and Broman, H. (1997). Signal separation applied to real world signals. In *Proc. of 1997 Int. Workshop on Acoustic Echo and Noise Control (IWAENC97)*, London UK.

Schafer, R. and Markel, J. (1979). *Speech analysis*. Wiley & Sons.

Schafer, R. and Rabiner, L. (1973). A digital signal processing approach to interpolation. *Proc. IEEE*, 61:692–720.

Schobben, D., Egelmeers, G., and Sommen, P. (1997). Efficient realization of the block frequency domain adaptive filter. In *Int. Conf. on Acoustics, Speech and Signal Processing (ICASSP)*, pages 2257–2260.

Schobben, D. and Sommen, P. (1996). Increasing beamsteering directions using polyphase decomposition. In *SSAP : Statistical Signal and Array Processing : IEEE signal processing workshop*, pages 117–120. IEEE Signal Processing Society.

Schobben, D. and Sommen, P. (1998a). A new convolutive blind signal separation algorithm based on second order statistics. In *Proc. Int. Conf. on Signal and Image Processing*, pages 564–569.

Schobben, D. and Sommen, P. (1998b). Realtime realization aspects of the cobliss blind signal separation algorithm. In *Proc. ProRISC/IEEE Benelux Workshop on Circuits, Systems, and Signal Processing*, pages 493–496.

Schobben, D. and Sommen, P. (1998c). Transparent communication. In *Proc. IEEE Benelux Signal Processing Chapter Symposium*, pages 171–174.

Schobben, D. and Sommen, P. (1999a). A new algorithm for joint blind signal separation and acoustic echo canceling. In *ISSPA 99 : Int. Symposium on Signal Processing and its Applications*.

Schobben, D. and Sommen, P. (1999b). On the indeterminacies of convolutive blind signal separation based on second order statistics. In *ISSPA 99 : Int. Symposium on Signal Processing and its Applications*.

Schobben, D., Torkkola, K., and Smaragdis, P. (1999). Evaluation of blind signal separation methods. In *Proc. Int. Workshop Independent Component Analysis and Blind Signal Separation*, pages 261–266.

Shimauchi, S., Haneda, Y., Makino, S., and Kaneda, Y. (1998). New configuration for a stereo echo canceler with nonlinear pre-processing. In *Int. Conf. on Acoustics, Speech and Signal Processing (ICASSP)*, pages 3685–3688.

Shimauchi, S., Makino, S., Haneda, Y., Nakagawa, A., and Sakauchi, S. (1999). A stereo echo canceler implemented using a stereo shaker and a duo-filter control system. In *Int. Conf. on Acoustics, Speech and Signal Processing (ICASSP)*.

Smaragdis, P. (1997). Efficient blind separation of convolved sound mixtures. In *Proc. of IEEE 1997 Workshop on Applications of Signal Processing to Audio and Acoustics*, Mohonk Mountain House, New Paltz, NY.

Smaragdis, P. (1998). Blind separation of convolved mixtures in the frequency domain. In *Int. Workshop on Independence & Artificial Neural Networks*.

Sommen, P. (1989). Partitioned frequency-domain adaptive filters. In *23nd Asilomar Conf. on Signals, Systems and Computers*, pages 677–681, Pacific Groove, CA.

Sommen, P. (1992). *Adaptive filtering methods*. Eindhoven (Netherlands): Eindhoven Technical University of Technology. Ph. D. Thesis.

Sommen, P. and Jayasinghe, J. (1988). On frequency domain adaptive filters using the overlap-add method. In *Proc. of the ISCAS conf.*, pages 27–30, Espoo, Finland.

Sondhi, M. (1967). An adaptive echo canceller. *Bell Syst. Tech. J.*, 46(3):497–511.

Sondhi, M., Morgan, D., and Hall, J. (1995). Stereophonic acoustic echo cancellation: An overview of the fundamental problem. *IEEE Signal Processing Letters*, 2(8):148–151.

Sorenson, H., Jones, D., Heideman, M., and Burrus, C. (1987). Real-valued fast fourier transform algorithms. *IEEE trans. on acoustics, speech and signal processing*, ASSP-35(6).

South, C., Horna, C., and Lewis, A. (1979). Adaptive filters to improve a loudspeaking telephone. *Electron. Letters*, 15(21):673–674.

Su, G. and Morf, M. (1983). The signal subspace approach for multiple wide-band emitter location. *IEEE Trans. on Acoustics, Speech and Signal Processing*, 31(6):1502–1522.

Tong, L., Liu, R., Soon, V., and Huang, Y. (1991). Indeterminacy and identifiability of blind identification. *IEEE Trans. on Circuits and Systems*, 38(5):499–509.

Torkkola, K. (1996). Blind separation of delayed sources based on information maximization. In *IEEE Workshop on Neural Networks for Signal Proc.*

Torkkola, K. (1997). Blind deconvolution, information maximization, and recursive filters. In *Proc. ICASSP*, pages 3301–3304, Munich, Germany.

Torkkola, K. (1999). Blind separation for audio signals - are we there yet? In *Proc. ICA and BSS*, Aussois, France. (Invited paper).

Turner, J. and Pretlove, A. (1991). *Acoustics for engineers*. MacMillan.

v. Gerven, S. (1996). *Adaptive noise cancellation and signal separation with applications to speech enhancement*. Katholieke Universiteit Leuven. Ph.D. Thesis.

Van Gerven, S. and Van Compernolle, D. (1995). Signal separation by symmetric adaptive decorrelation: Stability, convergence, and uniqueness. *IEEE Trans. on Signal Processing*, 43(7):1602–1612.

van Gerven, S., van Compernolle, D., Wauters, P., Verstraeten, W., Eneman, K., and Delaet, K. (1995). Multiple beam broadband beamforming : Filter design and real time implementation. In *Proc. IEEE Workshop on Applications of Signal Processing to Audio and Acoustics*, page 7.4, New Paltz, New York, U.S.A.

van Veen, B. and Buckley, K. (1988). Beamforming: A versatile approach to digital filtering. *AASP Magazine*, 5(2):4–24.

Wang, H. and Kaveh, M. (1985). Coherent signal-subspace processing for the detection and estimation of angles of arrival of multiple wideband sources. *IEEE Trans. on Acoustics, Speech and Signal Processing*, 33(4):823–831.

Weber, M. and Heisler, R. (1984). A frequency-domain algorithm for wideband coherent signal processing. *Journal of the Acoustical Society of America*, 76:1132–1144.

Weinstein, E., Feder, M., and Oppenheim, A. V. (1993). Multi-channel signal separation by decorrelation. *IEEE Trans. on Speech and Audio Processing*, 1(4):405–413.

Westner, A. (1998). *Object-based audio capture*. Massachusetts Institute of Technology. M.Sc. Thesis.

Widrow, B. and Stearns, S. (1985). *Adaptive Signal Processing*. New Jersey: Prentice Hall.

Yen, K.-C. and Zhao, Y. (1996). Robust automatic speech recognition using a multi-channel signal separation front end. In *Proc. 4th Int. Conf. on Spoken Language Processing (ICSLP'96)*, Philadelphia, PA.

Index

MIX
Papier aus verantwortungsvollen Quellen
Paper from responsible sources
FSC® C105338

If you have any concerns about our products,
you can contact us on
ProductSafety@springernature.com

In case Publisher is established outside the EU,
the EU authorized representative is:
Springer Nature Customer Service Center GmbH
Europaplatz 3, 69115 Heidelberg, Germany

Printed by Libri Plureos GmbH
in Hamburg, Germany